THE NAVIGATOR TRILOGY
BOOK THREE

THE
CUCKOO
PARADOX

DISCOVER WHY MIGRANT BIRDS CANNOT USE COMPASSES TO NAVIGATE

TED GERRARD

First published in 2015 by Samos Books
Broadford, Skye, Scotland IV49 9AQ
www.samosbooks.org

ISBN 978-0-9556439-2-7

Library Cataloguing in Publication Data

Gerrard, Edward

The Cuckoo Paradox

Key words
Bird evolution; Bird migration;
Bird eye pecten structure; Bird navigation;
Bird experiments; Cuckoo navigation;
Greenwich mean time; Ephemerides:

A catalogue record of this book is available from the British Library

Cover design by Edward and Anne Gerrard

Printing by Lightning Source UK Ltd., Milton Keynes MK13 8PR

TO MY WIFE

ANNE

CONTENTS

PREFACE

For the past six decades, tens of thousands of long-distance migrant birds have been trapped in nets and kept in aviaries prior to being placed individually in a variety of orientation test apparatus. Their frantic directional escape attempts are recorded, often for an hour or more, and the data from any individuals declining to co-operate is discarded. Only qualified behavioural experts are permitted to conduct such experiments and we are assured the birds are none the worse for the experience, although how anyone can know this has not been explained. The image of a robin with goggles taped to its head equipped with a clear foil over the right eye (and a frosted one over the left) during one such experiment does appear to suggest a degree of dishevelment.

www.cell.com/current-biology/ppt/S0960-9822(10)00779-7.ppt

As early as 1965 data acquired in test apparatus had, so one is led to believe, provided evidence that long-distance migrants possessed solar, stellar and/or magnetic compasses, and by inference could use these to navigate; to ascertain position, plot and follow a route.

This became known as the three compass method of navigation and was directly responsible for the following statement by David D Elliott, a member of the National Aeronautics and Space Council Executive Office of the President, when delivering the keynote address at *The Symposium on Animal Orientation and Navigation,* NASA Wallops Station, Wallops Island, Virginia in 1970.

"....the complex systems that animals use to navigate over great distances.......are marvels of micro-miniaturization and reliability, not yet duplicated electronically or mechanically."

It is not difficult to highlight the flaws in all these claims. Neither is it difficult to demonstrate the scientific impossibility of migrants being able to navigate long distances using all or any of these three compasses.

Conan Doyle, through his mouthpiece Sherlock Holmes, stated in *The Famous Four* "When you have eliminated the impossible, whatever remains, however improbable, must be the truth." But is there any alternative to the 3 compass method, however improbable?

My avian eye pecten structure hypothesis - an innate navigational system available to birds but not to mankind – appears to be the only candidate, but has been pointedly ignored for the past 40 years. How many readers have ever heard of a pecten structure, let alone its glare protecting properties?

Holmes might be able to get away with improbable solutions, but science rightly demands they be testable and one cannot test a truly untainted-by-human-hands long-distance migrant's directional responses to solar glare. In 1972 two talented visual neurobiologists, Horace Barlow (great-grandson of Charles Darwin) and Thomas Ostwald, did discover the shadow-casting properties of the pecten structure in a pigeon's eye; illustrated by a cheeky pigeon wearing a large peaked cap in a New Scientist review. Unfortunately they first had to kill the hapless bird.

In 2007, my first book in the Navigator Trilogy series –*Astronomical Minds; The True Longitude Story* – was based on the fact that Isaac Newton had invented the twin-mirrored, self-correcting angle-measuring device; the marine quadrant. The single most important development in the history of marine navigation. I constructed a copy using the same methods available to the Tower Mint in 1696 and demonstrated this in the National Maritime Museum at joint NNM/Royal Institute of Navigation conference in 2010. The design permitted the reflected image of one object to be overlaid onto the clear view of another and the real distance between the two measured accurately.

Viewing two objects simultaneously, one of which is protected by a filter at all but the very lowest of altitudes, mimics the avian eye twin pecten structures, a point that became apparent whilst testing my copy of Newton's instrument when the Sun was very low above the horizon (Chapter 4).

Three further pieces of supportive material have become available, this time from the ornithological experts themselves because of advances in satellite tracking systems.

The British Trust for Ornithology (BTO) fitted tracking devices to a number of adult European cuckoos and the birds' positional data indicated that they were not making use of the "3 C's" method. Then geolocator data obtained from small songbirds suggested likewise and finally two previously enthusiastic promoters of the avian magnetic compass reluctantly concluded that the *only* explanation to account for some recorded avian long-distance migratory navigation feats was that they possessed GPS devices.

Many of this book's diagrams were hand-drawn on a remote Hebridean island and originally published 45 or more years ago. These lack the precision of modern computer-based illustrations, but best represent an equally ancient hypothesis.

Ted Gerrard, Isle of Skye, 2015

11

PART ONE

A CONSPIRACY OF OPTIMISM

ELIMINATING THE IMPOSSIBLE

Chapter One

The Birth of the Avian Three Compass Concept

For many centuries the homing skills of pigeons and the autumnal disappearance and sudden spring return of all manner of bird species has been the source of much puzzlement. How does a bird come to possess such navigational prowess when many of us can become completely disoriented in a crowded car-park?

Although the search began in earnest in the 1920's, publication of key discoveries did not see light of day until the close of WW2 hostilities three decades later. Full references for all the following experiments can be found in Appendix 5.

Wild migrant birds used in orientation experiments.

A batch - a murder - of young hooded crows *(Corvus cornix)* were caught during spring northeasterly migration on their way back to their northern Baltic breeding grounds, and rather unfairly returned to their winter quarters. These then began their migration again, heading in same north-easterly direction. Thus the entire species, both adults and juveniles, were in possession of some kind of innate ability to migrate on a north-east/south-west axis. So far so good; migrants must possess an innate directional compass, one that is probably programmed differently for different species – NE/SW, N/S, and so on.

The details of an experiment that confirmed migrant birds were able to use the Sun as a directional compass were first published in 1949.

A test cage containing a captive starling had a perimeter ring of six equally-spaced windows that let in variable amounts of sunlight. Although it was soon realised that each bird would sensibly only attempt to escape towards a window, never a wall, the attempts were computed to the nearest eighth compass point. Phototactic escape responses were not considered. This explanation at best only fitted a species that migrated during daylight along a single directional axis.

Knowing from ringing returns and radar observations that many small long-distance migrants travelled at night and appeared to change their innate directions part way through their journeys, another professional ornithologist devised a new caged orientation experiment. This time to test whether migrants possessed a stellar compass, arranging at the same time to check whether his birds could adjust their innate direction part way through their journey. Details confirming this discovery were first published in 1956.

The operator had lain beneath the transparent floor of the tiny planetarium in order to make note of each directional hop of the single inmate. Only the most prominent of stars - no Moon or planets - were projected. When the positions of these prominent stars coincided with the latitude of the test rig at a given time of night, the inmate attempted to escape in a south-easterly direction – as expected.

But when the projection time and latitude was changed to represent the migratory mid-way point, the bird switched its major efforts towards the south – again as anticipated. The odd times at which each experiment was conducted appear to have coincided with the brightest stars being projected in the expected direction. Not easy for a third party to discover this back in 1956.

By now many ornithologists were convinced that migrants were born with innate solar and stellar compasses of some kind, that could be used to change direction part-way through their seasonal journeys. The two compass concept.

In 1958 the results of a lateral displacement experiment involving large numbers of free-flying starlings were published. The organizers hoped to show that navigationally inexperienced migrant starlings, when displaced *sideways* from Holland to Switzerland half way through their autumn journey from the Baltic to northern France/British Isles, would finish up in the wrong winter quarters in Southern France or even Spain. On the other hand adults would work out how to quickly compensate and arrive in northern France or the British Isles as per normal (Figure 1.1). Quite why the organizers of this very costly experiment should have *anticipated* that only *adults* might be able to reach their correct winter quarters after being shifted far to the east - across longitudes - and into unfamiliar territory, has never been explained.

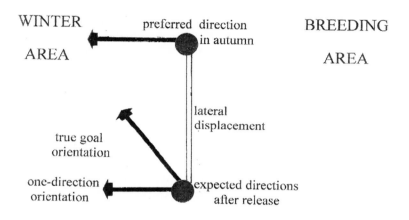

Figure 1.1 Outline of the Dutch starling displacement hypothesis.
Note the comment "expected directions after release".
Source: Derived from A.C.Perdeck (1958)
A different version published by Gerrard (1981a,b) is illustrated in Appendix 4.

Nevertheless subsequent ringing recoveries appeared to confirm the hypothesis. Adults, could indeed get back on course. Juveniles, on the other hand, could not, and continued on in their innate compass direction. In reality all but one single displaced adult also finished up in the wrong wintering area. Further details can be found in Appendix 4, and the full 37 page text of the experiment can be viewed at:-

http://www.nou.nu/ardea/ardea_show_article.php?nr=1562

Clearly long-distance avian migrants, however they manage, must "navigate" somehow, but the starling displacement claim introduced a touch of wonderment to the proceedings. Starlings can improve their innate navigational skills; they can actually *learn* how to get back on course after *lateral* displacement into unfamiliar territory. Wow!

These birds must therefore be in possession of an additional – third - compass, that would permit longitude to be assessed, but only *after* having experienced the migratory route. What could this possibly be?

Increasing numbers of migrants were now being recorded on radar screens whist flying beneath solid overcast at night. The magnetic compass concept might answer the displacement correction puzzle *and* flight under full overcast when the solar/stellar two compass method was inoperative.

At this point the hypothetical solution to the problem of using Earth's magnetic field to *learn* how to get back on course after lateral displacement was seriously tested. This 1965 experiment and the next two used an eight-sided cage with eight radial perches, all contained within an alterable magnetic field. Whenever a bird alighted on one of the perches, a directional hit was registered automatically. These were then bulked statistically. However the registration equipment did not indicate which way the bird was facing when it alighted on a perch.

First we were offered a cage with a choice of six windows and statistical evidence based on eight compass points, then came the apparatus with only a few bright stars to head for, then the seriously flawed free-flying claim and now a cage with two-way perches but only one-way records.

Five years later evidence was published showing that birds could detect variations in Earth's magnetic intensity and even assess its angle of dip at the surface.

But how do migrant birds acquire the information required to use a magnetic compass in the first place? Whereas this answer was the same as for Sun or solar compasses - they inherit it -, the magnetic compass claim carried slightly more conviction and offered larger scope for further research. This compass could be used day and night and in all weathers - more or less.

The inherited magnetic compass claim in turn raised the next question (for a second time). Different populations of the same species are known to migrate in different directions. The inherited information must therefore vary between groups. Does it? Yes, but the details were a long time in coming this time and were not published until 1989.

This experiment, and all the following used the Emlen test cage, (Figure 1.2) that consists of a funnel with sloping sides and a mesh top to stop the bird escaping. Each time the bird flutters up or down the sides of the funnel, its scratch marks are recorded. These are bulked statistically.

There are several problems associated with this type of test rig used in over 100 published claims – and still counting. For example, the birds are known to be attracted to imperfections - such as scratch marks inside the apparatus, and noise, heat and light variations outside.

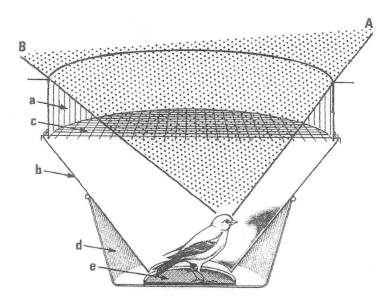

Figure 1.2 The Emlen test cage.

a. opaque circular screen. **b**. blotting paper funnel (later Tipp-Ex).
c. wire screen top. **d**. two quart pan. **e**. inky pad (later removed).
The projected lines of vision (A & B) have been added.
Source: Derived from S.T.Emlen and J.T.Emlen (1966)
Gerrard (1981a)

If different groups inherit different directional information, as they surely must, cross-breeding should provide the offspring with new navigational information. This was confirmed in 1992. So the magnetic compass is no longer just a pointer to the north magnetic pole, the avian owner of said compass has either inherited a magnetic declination chart - rather unlikely - or can re-calibrate its compass every so often with the aid its solar and stellar compasses.

Having seemingly solved the entire avian navigation puzzle - and somehow bypassing the question of how to *learn* how to get back on course - the question as to what actually operates the magnetic compass was raised and apparently answered in 1994. Melatonin in the bird's eye is the key. The crucial melatonin discovery was further refined in 2002.

So a string of ten questions, raised one by one in a logical sequence, produced explanations obtained from caged experiments backed by doubtful experimental proof, one by one in the same sequence. These claims incidentally all stemmed from German research.

This covers all the original basic experimental claims conducted in various test rigs with the exception of experiments in the United States which revealed that nocturnal migrants when tested under planetarium skies could not take up the correct orientation - thus confounding the original claim of a "genetic star map". Instead it was suggested *"a maturation process in which the stellar cues come to be associated with a directional reference system provided by the axis of celestial rotation."*

This was how the concept that long-distance migrants are able to navigate from point A to point B was formulated. Each group of birds obviously inheriting a different series of compass bearings and timings. *Juveniles* possess an innate "fixed" compass direction (solar, stellar, magnetic or any combination of the three) that could be maintained for each sector of any migration route and then changed for the next section. On the other hand, *adults* could additionally compensate for lateral displacement, even into unknown territory.

Chapter Two

The Flaws in the Avian Three Compass Concept

Orientation versus Navigation; Longitude & Mean Time;
The Cuckoo Paradox; Advance Prediction Tables.

None of the claims discussed in Chapter 1 appear to have been successfully replicated under controlled conditions, possibly because there are several underlying flaws in the concept of compass-guided long-distance avian migratory navigation.

When discussing avian long-distance migration, the term "compass" refers to any mechanism that can be used to select and then maintain a heading.

For example. The avian solar compass. By being able to tell the time of day by observing the Sun's position in the sky, the migrant bird can use this information to travel in a chosen direction, subsequently maintaining that direction by constantly updating the changing position of the Sun. Or when the Sun is masked by cloud, possibly using polarized light in a similar manner.

The avian stellar compass. By being able to do likewise with stars or star patterns at night.

The avian magnetic compass. By determining the position of magnetic north and using this information to travel in *some other direction*. Subsequently maintaining that direction by regularly updating the constantly changing apparent position of magnetic north.

Flaw number one. Orientation versus navigation.
Defining the words *orientation* and *navigation* in the context of long-distance two-way avian migration now require an explanation.

If a captive bird attempts escape by *repeatedly* heading towards a lighted window, or from a cage, that is an orientation event. It matters not whether the escape direction happens to more or less match the migratory direction that the escapee might then adopt for its migratory flight; it is still an orientation event and no proof of a navigational ability.

If the Sun, a star, scratch marks, directional noise or anything else happens coincidentally to be the attraction or distraction, the inmate is still only responding to an *orientation cue* in a "fight or flight" manner. It is not necessarily attempting to navigate, nor is it proof that it is in possession of a solar, stellar or magnetic compass of some kind.

A racing pigeon, when released in Spain from its basket that is already lined up facing towards its distant home in Belgium, naturally has to escape in the direction of home, like it or no. That response is an *orientation* event and has nothing to do with navigation. In exactly the same way as it would be an orientation event if its basket was lined up in some other direction. But if that pigeon then, after "getting its bearings" heads off directly to its home loft in Belgium *and arrives there*, that is a feat of *navigation*. Whether it flew entirely across familiar ground or followed others, or used some other means, it was still an act of navigation.

The organisers of the caged experiments already outlined, were claiming evidence of migratory navigational abilities whilst simply recording whole rafts of unsuccessful escape attempts, and the early Dutch starling free-flying displacement case conveniently gave credence to those caged claims because migrant birds could apparently hone their navigational talents by experience.

Flaw number two. Longitude and Mean time.
Let us assume for a moment, that orientation does equal navigation and that all the caged experimental claims mentioned in the previous chapter have been confirmed by independent research.

The mandatory **Longitude.**

Without being able to determine longitude, long-distance sailors or aviators could not correct for lateral displacement, caused for example by crosswinds whilst crossing an ocean or featureless desert.

A (meridional) line of longitude is an imaginary line drawn from pole to pole and represents the shortest distance between you and the equator, wherever you happen to be. Take a few steps to one side and you are now standing on a *different* imaginary line of longitude. Even now one cannot walk directly "down" or "up" a line of longitude - let alone fly along one - without using continually updating information from orbiting satellites,

The mandatory **Mean time**.
Aristotle believed Greek swallows hibernated in the mud at the bottom of lakes; he also put it about - with some justification - that we lived on a stationary perfect sphere. Actually we live on a tilted spinning oblate spheroid; which makes for extremely difficult timekeeping.

Earth's annual orbit round the Sun is elliptical and constantly changing speed, so the time at which the Sun reaches its highest point in the sky each day (local noon) varies from one day to the next. The *average* value of all the constantly changing *solar* days in each year is termed a *mean solar* day and it is this average value that is used to set our timekeepers to; Greenwich Mean Time for example.

If we kept time to coincide with local noon, our church clock would require altering daily. So registering *mean* time at the point of departure is essential for anyone wishing to make use of solar or stellar prediction tables (ephemerides/almanacs) that inform us in advance, of the position of any solar or stellar object on any given date.

Incidentally the circadian clocks that regulate all manner of internal responses, human as well as avian, work on local time, *not* GMT. This is why we suffer more from jet lag after flying east. Thus the migrant would have to possess two timers, not one.

The principle of retaining GMT (or UTC–Universal Constant Time), throughout a long journey is how the tiny geolocators fitted to avian migrants can provide details of the birds' en-route positions – if and when they can be re-trapped and the data downloaded. This is done by recording the times of sunrise and sunset and day length en-route on a memory chip. Because the latitude and *longitude* of original fitting location is known through the auspices of GPS, it works well for much of the year and at most latitudes. This timing system also serves to keep satellites in orbit and our in-car GPS systems reliable.

So it is not possible to use the Sun or stars to navigate across trackless wastes from A to B without knowing the precise location of both places, and without possessing an accurate reliable timer set to the *mean* (average) time of the departure longitude. But one cannot determine the *mean* time without setting up an observatory at point A in order to produce almanacs, and one cannot make use of those complex tables without knowing the departure date.

Obviously each departing long-distance migrant cannot construct its own observatory and cannot hang around for years whilst compiling the data, cannot possess a timer set to mean time, and cannot be aware of the essential departure date to make sense of said data. So into the trash basket go the solar and stellar compasses; again.

To make use solely of a magnetic compass is also impossible without foreknowledge of A and B and the annual global magnetic declination changes and local deviation values.

Flaw number three. The cuckoo paradox.
The most inexplicable of all avian migrations are performed by some cuckoo and other parasitic species. In the case of the common cuckoo, the hen bird lays many eggs singly in different nests and is promiscuous. The unwitting foster parents, having had one of their own eggs removed by the laying cuckoo and the rest later ejected by their foster child, then exhaust themselves in feeding this rapidly growing assassin; by which time the parent cuckoos have sensibly departed for warmer climes in central Africa, probably south of the equator.

The foster parents, if they are themselves of a migratory inclination, then also leave, but not for the same African areas. Thousands of young cuckoos are abandoned, but if they are to survive, they too will have to migrate to warmer climes well before the onset of winter. No one to tell each and every scattered one of them where they are or where they have to go to; no marked map (let alone a GPS/UTC timer) in their in-flight travel bags. How do they succeed in finding their way from one unknown location to another? Joining up with other clueless youngsters of the same ilk is unlikely to be of much help.

However when the fairly representative track of an *adult* common (European) cuckoo *(Cuculus canorus)*, is examined, it is abundantly clear that even adults, if miraculously they were to somehow possess "3 C" abilities and all the ancillary baggage that has to go with this, are *not making use of any of it*.

http://www.bto.org/science/migration/tracking-studies/cuckoo-tracking

Observe the migratory tracks of these adults, the crossing of previous paths, the back-tracking, the long rest-up periods, the lateral movements and the sudden long non-stop flights ....and ponder.

But those young cuckoos, who cannot possibly know the geographic location of their departure point or the precise *calendar* date of lift-off, and cannot possibly possess a clock set to departure longitude *mean* time, can at best **leave home and head elsewhere**. Adults of the species may well possess prior knowledge of both breeding and wintering areas, but not those youngsters.

Logic suggests that if this cuckoo paradox can be explained, every lesser avian navigational feat should fit within that same framework. It matters not what that explanation might be, just so long as it is scientifically sound.

However, before offering up a possible explanation for how a young cuckoo can unwittingly fly from Scotland to central Africa without the aid of the "3 C's", it is important to highlight how long it took mankind to work out how to get from one *known* location to another. What a rocky branching path littered with dead ends, blockages, lucky guesses and highway robbers that was. No simple steps, one after another, all in the right direction for *Homo sapiens* during that investigation, but tens of thousands of years of trial and error.

Only one species on this planet can read, write and pass on acquired data from one generation to the next and reach conclusion from such evidence. Only thus can that single species *predict* the positions of stellar objects or Earth's magnetic declination, or manufacture timers capable of registering and holding *mean* time, publish ephemerides and then use such equipment to *navigate* over long distances. The collective ingenuity of countless generations finally, in 1776 permitted *Homo sapiens* to do so. The key discovery being how to determine longitude *whilst on the high seas* (Chapter 4).

Chapter Three

Establishing the Surface Dimensions of our Planet

How could birds know all there is to know about navigating across the surface of a spinning oblate spheroid millions of years before we managed to work it out?

Living on a plate.
Until around 600 BC those who had the time or inclination to think about it, believed Earth was a flat plate and sailors venturing out of sight of land risked falling off the edge. Early Greek philosophers believed that this world (somehow surrounded by the river Oceanus) on which they lived was at the very centre of a huge hollow sphere. Above was the vault of Heaven on which all the heavenly bodies were glued and below was Hades. The flat fiery disc of the Sun rose in the morning from Oceanus in the east, flew across the sky under the dome of Heaven, and splashed down again in the far west every evening. How it got round to the east again ready for the next dash across the firmament was obviously in the laps of the gods and beyond the bounds of discussion.

If not a plate, how big is the sphere?
Pythagoras (circa 570-495 BC), born on the Aegean island of Samos, was probably one of the first to suggest that Earth was a sphere rather than a flat plate. This would have been in keeping with his belief that the sphere was the most beautiful of solid figures. Possibly because of his rumoured preferences for eating little else but beans, as well as making a general political nuisance of himself, his ideas and discoveries were only properly publicised after his death.

Aristotle, (384–322 BC), tutor to Alexander the Great, mentioned that mariners had noticed that familiar stars that were only ever seen on the northern *horizon* when leaving Alexandria, were higher in the sky on arriving in Cyprus or Rhodes. The difference in angular height of these stars between Alexandria and Rhodes led him (or associates) to conclude that Earth's circumference was 400,000 stades and that the diameter of the Sun was considerably larger. Stades were ancient Greek or Egyptian measurements of length, originally of sports stadia. One Royal stade (RS) was equal to 210 or 211 metres and Egyptian sporting venues, being for some reason somewhat shorter, were equal to 157.5 metres (ES).

Enter Eratosthenes (276-194 BC), keeper of the great library at Alexandria, which by his time contained hundreds of thousands of scrolls, many collected by Aristotle. Eratosthenes had heard travellers' tales that the Sun shone directly down a well at Syene in southern Egypt on midsummer's day. In a flash of pure genius he realised this peculiar event could be used to measure Earth's polar circumference.

He measured the angular height of the Sun at noon on midsummer's day at Alexandria and discovered it was 1/50th less than the vertical at Syene (82.8 degrees rather than 90). Guessing the distance between the two points was 5,000 ES, and hoping they were on the same meridian, he multiplied 50 by 5,000 to arrive at Earth's polar circumference of 250,000 ES (39,350 km.).

The distance between Syene and Alexandria is 5,200 ES and they are not on the same meridian but if they had been, the distance would indeed have been 5,000 ES. The well at Syene was not exactly beneath the midsummer's day noon Sun but Eratosthenes' error in measuring the height of said Sun at Alexandria almost made amends. Thus these four very excusable errors somehow almost cancelled each other out.

Unaware that he had arrived at a figure only some 2% removed from reality by pure luck, he realised that if he was going to split the surface of Earth into horizontal hoops for mapping location purposes, each "degree" (hoop) would equal 694 and 111/250th's ES. Not a good idea, so he simply added 2,000 ES to his estimate, making 252,000 for the circumference and 700 for each "degree" of latitude. This final figure, equal to 39,690 km., was actually now within 1% of Earth's true polar circumference of 40,008 km.

Ptolemy (circa 85-165 AD) also lived in Alexandria, where he made astronomical observations between 127 and 141 AD. Although aware of the work of Eratosthenes, he, for some inexplicable reason, preferred to work with 500 stades (of some sort) per degree of latitude for his global mapping project.

Over the centuries, Ptolemy's updated maps and charts and various copies of his *Geography* were translated into a number of languages and distances, all or any of which were snapped up by mariners and explorers. If not, a copy of a copy - one immediate area for further error of course. Some translators of Ptolemy's work became further confused over the length of the two different stade measures when converting into Arabic miles (of about 1,926 metres each) or Roman miles (about 1,478 metres).

Is our planet stationary or spinning in orbit round the Sun?

A theoretical method of fixing one's position with the aid of the changing locations of stellar objects had been known for thousands if not tens of thousands of years, but turning theory into practice was hamstrung by an absence of reliable prediction tables, timers that could retain local time at the point of departure, and accurate measuring devices and charts. But there had always been a larger fly in the ointment – geocentricity.

Of course it's stationary.

Living on the surface of a large stationary sphere at the very centre of the Universe, with Sun, Moon and other stellar objects all completing orbits daily, seemed logical to the northern hemisphere inteligencia, even after Pythagoras had put paid to the flat plate notion. Despite the puzzle over how people in distant lands managed to stay upright on a large ball, Earth remained firmly static; and why not? If it was as large as was being claimed, it most certainly could not be spinning; where were the enormously powerful directional winds that would surely blow everyone over if it was Earth that spun once daily?

But to dismiss the highly unlikely spinning Earth theory out of hand in favour of a Sun zipping round a stationary Earth at enormous speed just because the streets of Athens were often windless, was, had anyone realised, simply asking for trouble.

For example, even before the flat plate notion had been scuttled, Anaximander of Miletus (circa 610-546 BC) had been of the opinion that transparent concentric spheres carried heavenly bodies round Earth. This originally only involved the obvious; the Sun and Moon plus all the thousands of pinpricks of light in the night sky. Anaximander thought that the stars (which he lumped together as a single group) were actually nearer to his earthly observation platform than the Moon.

Just to confuse matters, Heraclitus (circa 535-475 BC) of Ephesus to the north of Miletus - a city on the Adriatic coast of Turkey - announced that a new Sun was manufactured each morning by the sea in the east vaporising, the vapour rising and changing into air and fire. The fire turned into the Sun that, although only one foot in diameter, shone brighter than the Moon because it was moving through cleaner air. In the evening the Sun was sucked towards the sea in the west. The fire of the Sun condensed back into water and the water into earth. A sort of reverse osmosis – and so on, day after day.

This fond belief was to be the death of him. Suffering from a water-retention ailment, he buried himself in a pile of farmyard manure hoping that the heat from the dung heap would convert his excess water into vapour in accordance with his hypothesis. The fact that this did not work may have undermined his reputation, although he does seem to have found his way into *The Guinness Book of Records* for claiming the Sun was the size of a cabbage.

Eudoxus (circa 406-355 BC) was yet another remarkable man to emerge from that tiny area of the eastern Mediterranean, in this instance some two day's ride south of Miletus. By now advances in astronomy had properly identified 5 planets; Mercury, Venus, Mars, Jupiter and Saturn. Not for a moment did it occur to anyone that they were actually living on a 6th!

A celebrated geometer, physician and legislator, Eudoxus attempted to bring some semblance of mathematical order to Anaximander's concentric spheres hypothesis, because this arrangement could not account for the apparent varying diameter of the Moon, the changing luminosity of at least one of the planets, and alarming changes of direction of Mercury and Venus.

His model Universe contained spheres stuck on spheres that carried a planet stuck at an angle – and so on, and some of the large concentric spheres imparted motion to other large spheres. He forcefully promoted his updated version that consisted of no less than 27 concentric spheres and spheres within spheres and equally forcefully promoted the "all heavenly motions are circular and move at constant speeds" dogma. He insisted that if the facts suggested otherwise, the facts were wrong! All utter nonsense of course, and also like many other illogical scientific hypotheses, very difficult to combat because so very few could comprehend the details.

At the primitive level of marine navigational expertise that existed 2,000 years ago, it did not matter much whether Earth was stationary and everything else buzzed round it or vice versa. But as more and more hollow spheres had to be added to account for new astronomical discoveries, matters became so complex that the production of half-way reliable solar or stellar prediction tables remained unattainable.

A Heavenly clock?
In Egypt and Mesopotamia as far back as 1,500 BC, sundials (day) and water and candle clocks (night and day) were already divided into 12 hour sectors and could probably register local time to better than 10 minutes. Sundials might at a pinch provide a fair idea of latitude but none of these timers were good enough for oceanic navigation purposes. Oddly astronomers had been aware of a possible solution to this problem since about 130 BC.

Way back in time, elders had meticulously noted the days of full Moons, summer and winter solstices and both equinoxes. These key annual obvious occurrences were easy to recall, provided one had a few pieces of heavy equipment strategically placed - Stonehenge being an early example. This saved trying to remember when to plant crops or sacrifice virgins usefully, but did little to solve the puzzle of the Moon's erratic path through the heavens.
Then a bright spark who had been checking through old clay tablets or fragments of parchment, realised that the lunar monthly cycle matched with an annual calendar date only every 19 years. For example a full Moon at the summer solstice date did not occur again until 19 years

later. Actually the "bright spark" was an Athenian 5th century BC astronomer by the name of Meton who had checked his figures by erecting pillars and noting on them the solstices, but was not linked with this discovery in his lifetime. His assistant Euctemon seems to have taken the credit and much good did it do him because the figures were wrong.

There were almost exactly 235 lunar cycles in 19 years (of 365¼ days each) but as records became more reliable it was realised that a 4 x 19 year cycle was a better fit. Divide the number of days in 76 years (365¼ x 76) by the number of lunar months in 76 years (235 x 4) and one lunar month equals the Greek fractional equivalent of 29 days, 12 hours, 44 minutes and 25½ seconds. But still not an exact match. As the life span of any one enthusiastic recorder was unlikely to exceed any further multiples of this cycle, by then known as the Callippic 76 year cycle, they were stumped.

The astronomer Hipparchus was born in what is now Northern Turkey in 190 BC. He worked for many years in the city of Alexandria and also set up an observatory on the island of Rhodes, where he died 70 years later. He was one of the first to realise that the slow nightly slippage of the Moon's position in relation to the backdrop of stars could be used to assess longitude if only this could be predicted. An absolutely brilliant concept.

Hipparchus decided that the Callippic 76 year cycle could only yield a closer approximation if upgraded further still. So he multiplied this 76 year cycle by 4 again, making a 304 year Hipparchus cycle. Then, knowing the year length of 365¼ days was almost certainly just a shade too long, he knocked off the least possible amount; 1 day in 304 years. This had the effect of reducing the year length by nearly 5 minutes and as a consequence the lunar month time came down in sympathy by 23 seconds to 29 days, 12 hours, 44 minutes, 2½ seconds. Within less than a single second of the correct figure, and all achieved without a clock. Although there were 2 minor errors in his calculations, these conveniently cancelled each other exactly. Eratosthenes was not the only lucky Alexandrian astronomer.

Hipparchus then spent the rest of his life - about 5 years - unsuccessfully trying to predict the Moon's nightly position in advance.

Not at the centre of the Universe after all ?

The geocentric belief persisted - despite Aristarchus of Samos having advanced an heliocentric hypothesis 250 years before the birth of Christ - until Nicholas Copernicus read about this in the Vatican library. He then published an updated version in 1543. Even so, perfect symmetry was retained, which meant that the Sun had to orbit a central point annually.

The (re)invention of the telescope in 1608 by the Dutchman Hans Lippershey had enabled Galileo Galilei, a little more than a year later, to study the planet Venus in detail, knowing that if he could observe its changing shape this would provide conclusive evidence that Earth orbited the Sun. He was so confident that he would succeed, (and more to the point worried that someone else would beat him to it) that he sent the following message to a few of his trusted supporters *before* Venus was in position to provide the evidence:-

Heac immatura a me iam frustra leguntur o.y - These are at present too young to be read by me. "So what" was probably the likely response and no one realised it was actually a cryptic message.

By early 1611 Venus had moved to one side of the Sun and was exhibiting a crescent phase; proof produced and fame secured, Galileo revealed the unscrambled version:-

Cynthiae figuras aemulatur mater amorum. - The mother of love (Venus) imitates the shape of Cynthia (the Moon).

Galileo had used abbreviations (*o.y.* - by me) in his first message to make a perfect fit for the second and this had allowed him to construct dual messages *from the one string of 35 letters*. He must have devoted long hours to the composing of his hidden message that could only be recognised as such by someone with an intimate knowledge of the sender and subject. Importantly, he had hedged his bets. If no crescent phase had been observed he would, presumably, not have revealed the coded message. He used a similar method in announcing the discovery of Saturn's rings and on that occasion confused his most ardent supporter Johannes Kepler into thinking he had discovered moons orbiting Mars.

Only when Kepler (1571-1630) produced proof of the elliptical orbit of Mars and a scientific explanation for planetary elliptical orbits - which in turn allowed for the central Sun to remain static - did it become possible to consider working on the production of reliable prediction tables.

Jupiter's moons – a second heavenly clock?
Galileo's version of Lippershey's telescope also enabled him to discover four tiny moons orbiting the planet Jupiter. These buzzed round the planet at predictable speeds, sometimes disappearing behind Jupiter and then reappearing again on the other side as if by magic. It became painfully obvious, even to those who were unwilling to accept the evidence provided by the changing outline of Venus (evidence of a Sun-centred system), that at least 4 of God's creations were misbehaving. Galileo's earth-shattering discovery was rudely dismissed by the Jesuit opposition thus:- *"There is no proof that anything seen viewed through these curved glasses exists anywhere except in those lenses. This is because what is seen disappears when the lenses are removed."*

Importantly, Kepler and Galileo had both realised that the "stars" of Jupiter could be used for the same purpose the Moon might eventually be used; as the hands of a heavenly clock in order to determine longitude. When one of Jupiter's moons suddenly disappeared from view behind or in front of its mother planet, that event would be observed *at almost exactly the same moment wherever the observer was located on Earth's surface.* Check the time of local noon, compare this with the your tables that told you the time that event was occurring at Greenwich or wherever, and the difference in minutes would equal the number of ¼ degrees of longitude west or east from Greenwich the observer was. This heavenly clock would surely prove far easier to predict than the motions of the Moon and both astronomers were absolutely right.

A Renaissance Comedy – how big did those ancient Greeks say it was?

What had been a fairly accurate assessment of the actual distance of a degree of latitude at the time of Eratosthenes, had become a complete muddle 1,300 years later. A muddle based on hearsay and not on any scientific evidence.

For example, by the close of the 15th century, the Portuguese had settled on 16 2/3 leagues per degree, 1 league being equal to somewhere between 5,920 and 5,926 metres. About 98.7 km., or 35,570 km. for the polar circumference and erring by some 12% on the small side. By 1503, 17½ leagues had been substituted; a major step in the right direction (103.7 km. and 37,336 km.) but still nearly 7% less than the true figures.

In 1635 an English surveyor, Richard Norwood (circa 1590–1675), measured Earth's polar circumference by the Eratosthenes method. But he used legwork and a magnetic compass instead of guesswork, and a surveyor's chain instead of the grapevine. Norwood's estimate for Earth's polar circumference was 40,291 km., or about 0.6% too much. Eratosthenes' lucky figures all those centuries earlier had been an underestimate of about 0.7%. This ex-apprentice fishmonger's book, *The Sea-Man's Practice,* published in 1637, can be found in an even greater library than that of Alexandria; the British Library in London.

It was left to the French to sort the matter out in a true scientific manner once and for all – once again, almost. In 1669, the Académie Royal des Sciences issued instructions to the astronomer and surveyor Jean Picard to establish the circumpolar circumference of Earth by the triangulation method. Picard and his august group, with the further assistance of two pendulum clocks and a number of cumbersome telescopes fitted with cross-hairs, first established two base sites on the same meridian by the Jovian moon method (see Chapter 4). From these sites near Paris and Amiens they then set up 13 great triangulation points by which they measured the total distance between the two base sites without having to measure every step of the way. The latitudinal difference between the base points was 1 degree, 11 minutes and 57 seconds; the distance 68,425 toise (at 1.949 metres to a toise, 133.37 km.), making one degree of latitude equal to 111.22 km., – almost exactly correct – *for the latitudes involved.* This in turn eventually produced proof that Earth was not a perfect Pythagorean sphere after all – it was fatter round its middle. Not everyone was pleased! Some years later when the map of France was slimmed down as a result of Picard's painstaking work, Louis XIV who was also somewhat fat of waist, famously pointed out that he had lost more territory to his astronomers than to his enemies.

Length of a degree of latitude and of Earth's polar circumference

Actual measurements	**111.33 km.av./40,008km.**
Aristotle (Royal stades)	**233.33 km/84.000 km.**
Aristotle (or Egyptian stades)	**175.0 km/63,000 km.**
Eratosthenes (Original)	**109.3 km/39,350 km.**
Eratosthenes (after additions)	**110.25 km/39,690 km.**
XV Century Portuguese (aprox.)	**98.7 km/35,570 km.**
XVI Century Portuguese (aprox.)	**103.7 km/37,336 km.**
Norwood	**111.92 km/40,291 km.**
Picard	**111.22 km.**

It had taken some of the world's best brains two thousand years just to produce a fair estimate of Earth's polar circumference and to make the disturbing discovery that it was not a perfect sphere. We now also knew that it was spinning once daily as it rushed at a frightening pace in an elliptical orbit round the Sun. Yet we could only hazard a guess as to where on our globe Japan was in relation to Lisbon. At least now *Homo sapiens* had the ammunition with which to seriously address both parts of the navigational conundrum; how to get from the known position of point A to the known position of point B, and only then how to plot ones position anywhere on the planet.

Several great mathematical brains set to work on the first part of this puzzle more or less simultaneously following the completion of the Paris Observatory in 1667 and Picard's published results 2 years later.

Chapter Four

Solving the Problem of High Seas Navigation

Discovering the relative positions of A and B.
In 1674 the brilliant mathematician John Gregory, whilst attempting to obtain funding for a Scottish observatory at St. Andrews University, was the first to mark out an actual meridional line; across the floor of his workshop. That same year he wrote to Jean-Dominique Cassini, the Italian-born director of the Paris Observatory, requesting data on the partial solar eclipse of August 12th 1673. Data that would possibly enable him to determine the position of St. Andrews in relation to that of Paris. It seems that he received no reply.

Following intense lobbying from Sir Jonas Moore, Surveyor General of the Ordnance, and Sir Christopher Wren (both also able mathematicians), Charles II gave permission - but no funds - to construct an observatory in Greenwich Park. This was hastily completed, but ill equipped by 1676. The opening on June 6th timed to coincide with a solar eclipse was attended by the monarch, who arrived too late to observe the less than spectacular events; the poor view of the partial eclipse and the equally poor construction of the observatory.

Eventually the positions of London and Paris were properly affixed to world maps, courtesy a combination of timed eclipses and the Jovian moons method. The prime meridian becoming Paris if one was French and Greenwich if one was English. But it was not all plain sailing because an accurate timer had been required in order to produce the Jovian ephemerides. The dispute over who invented the timer did nothing to improve international relations.

In or about 1656, the Dutch mathematician Christiaan Huygens had invented the pendulum clock and 10 years later became the director of the Académie des Sciences in Paris. So far so good and pendulum clocks had been responsible for the accurate placement of both capitals.

The ones in Paris made by Huygens and those at Greenwich by Thomas Tompion, a colleague of Robert Hooke. Discovering the relative positions of A and B had been solved.

Discovering the position of B when only that of A is known.
This should have been relatively straightforward – a logical progression based on the same Jovian moons' method.

As early as 1670 Huygens, working for the French, had claimed he was well on the way to perfecting a sea-going pendulum clock. Then suddenly, French research into marine pendulum-driven weight-powered clocks was shelved and Huygens had switched his attention to spring-powered balance-wheel regulated watches. He had announced this switch to Royal Society colleagues in 1675, in a Latin cypher, the English version of which is:-
aaaaaabcccccdeeeeeeeeffhhhhiiiiilllmnnnnoooooprrrrsssttttttttvx

As with the Galileo cypher, not a lot of use to anyone, but at least this one was recognisable as such. The following month Huygens had revealed an unscrambled version, that had upset Hooke who promptly accused Huygens of stealing his idea. To all but Hooke, the Huygens plain version was almost as indecipherable as the encrypted one!
The axis of the movable circle is attached to the centre of an iron spiral.

The unfortunate Huygens had been forced to switch his research towards watches with wind-up mechanisms and balance springs because it had been discovered that pendulum clocks lost time in lower latitudes. Isaac Newton was later to correctly attribute this to Earth's equatorial buldge.

Sadly it also meant that free exchange of information between fellows of the Royal Society was compromised. Cassini had received his fellowship in 1672, Hook, Wren and Huygens were already founder members and Flamstead became a fellow following the comptetion of Greenwich Observatory.

Lacking access to Cassini's data, Flamsteed published a number of inaccurate tables in England between 1683 and 1686. He was in such a hurry to publish his Jovian almanac for 1684 in the Royal Society's

Philosophical Transactions, that he never even corrected the proofs, thus confusing the issue somewhat; *"A letter from Mr. Flamsteed concerning the Eclipses of Saturn's Satellit's for the year following. 1684 with a Catalogue of them, and informations concerning its use."*

What mariners made of the title's typographical mistake - Saturn instead of Jupiter - when they were hard put to locate either, is anyone's guess. What they thought of Flamsteed's sub-text is sadly all too obvious. *"And I must confess it is some part of my design, to make our more knowing Seamen ashamed of that refuge of Ignorance, their Idle and Impudent assertion that the longitude is not to be found, by offering them an expedient that will assuredly afford it, if their Ignorance, Sloth, Covetousness, or Ill-nature, forbid them not to make use of what is proposed."*

Edmond Halley, at this time the official Clerk to the RS, much to Flamsteed's annoyance then arranged for the RS to publish Cassini's far more accurate *"New and Exact Tables for the eclipses of the First Satellite* (Io) *of Jupiter"* in 1694. But Halley, like nearly everyone else bar landlubbers Flamsteed, Hooke, and Wren realised that using the Jupiter moons' method of determining longitude on the high seas would be impossible – and he was right. At least the way was now open for the fixing the positions of other locations, however distant on that world map, provided that the fixer could get home to compare his data with that of his master observatory.

Linking a long sea passage to the position of master observatories in Paris or London should have been a priority for any mariner with access to the data of one or the other. Strangely it was not the case, but to understand why not, we have to go back in time once again.

All mariners, especially the Portuguese and Spanish, were aware of the importance of establishing an easily recognisable, static location from which dead reckoning could commence; a prime meridian. The most westerly of the Portuguese owned Cape Verde islands for the Portuguese and Mount Teide, on the Spanish owned island of Tenerife for everyone.

Mount Teide soon usurped Cape Verde because the snowy peak of Teide could be sighted from upwards of 100 miles without having to close it. In reality Teide remained the prime meridian on Atlantic sea charts even after functional observatories had been constructed, because so many seafaring nations were denied access to those observatories data.

As recently as the beginning of the 18th century Nathaniel Colson's 160 pages of marine tables *(The Mariners New Kalendar)* still provided longitudes west and east of Teide. Admiral Sir Cloudesley Shovell's fleet navigators were also using the *Kalendar* data at the time of the disaster in 1707, and had they paid attention to Halley's published data, some 1,800 lives would have been saved.

On the other hand William Dampier's map *(A Voyage to New Holland p 35)* published in 1709, depicts Lizard point as his prime meridian, years after Halley had based his three *Paramore* voyages and his chart of the magnetic variation over the Atlantic on the Greenwich meridian.

It made sense for Halley to use his departure point at Greenwich as the prime meridian. Equally it made sense for Dampier, a sea dog of the old school, who had already circumnavigated the globe, to use the last bit of terra firma he was reasonably sure of.

Halley, because of his intimate astronomical knowledge and because he possessed an accurate angle-measuring device (see below) would be the first to determine the longitude of Barbados (Jupiter eclipse of Io, 24th April 1699), and on an illegal visit to the Portuguese island of Sal, Cape Verde (Jupiter transit of Ganymede 2nd November 1699).

Discovering ones position on the high seas.
So by 1700, about the only global navigational problem still outstanding was that of determining one's position on the high seas; determining both latitude and longitude. Surprisingly whilst longitude had always been the stumbling block, latitude was still almost impossible to determine accurately other than in exceptionally calm conditions.

All very well to know where London was in relation to York or Paris or even to Barbados or Sal, but mariners, with the single exception of

Halley, still could not determine *latitude* accurately on the high seas. *Latitude*, not *longitude* - achieving both was still far in the future.

The marine self-correcting octant.

Although angle-measuring devices have been available to assess the height of heavenly bodies from day one of our existence (an upright stick casting a shadow), those suitable for marine use were probably first used by unknown explorers using an upright mast or spar to check on when a home star really was directly overhead. Later the astronomers' large fixed stone quadrants plus movable arm with 2 pinhole sights and primitive divisions, were adapted for other angle-measuring tasks. After all, surveyors and map-makers could not be expected to dismantle a stone dioptre, pack it on the backs of camels and re-assemble it in a far off city just to determine latitude.

Eventually a small wooden or ivory portable version was produced and this in turn developed into the mariner's quadrant hung from the rigging, with a plumb line being employed to display the observed angle. Then came the brass astrolabe, the hand-held cross-staff and, by the late 1600's, the back-staff or Davis quadrant.

With the improvement in the accuracy of solar and stellar ephemerides, came the demand for more accurate and reliable marine angle-measuring devices.

The invention in 1695 or 1696 of the marine all-weather self-correcting angle-measuring device by Newton was the result of another of his eureka moments that he had been so good at plucking out of thin air. On this occasion, Newton, by then Warden of the Tower Mint had been egged on by Secretary to the Admiralty Samuel Pepys (..."*discover some method of ascertaining longitude at sea*"). Result – the single most important invention in the annals of maritime navigation; assisted it must be pointed out, by the practical expertise of Halley.

Halley, during his voyages of discovery as commander of the naval vessel *Paramore,* became the first person ever to determine *latitude* accurately on the high seas often to within 2 or 3 minutes of arc. But only because he had Newton's precious marine octant on board and possessed his own personally calculated ephemerides.

The magnetic compass.

Because Halley could determine his latitude accurately, he could make good use of a magnetic compass to determine magnetic variation across the entire length and breadth of the Atlantic Ocean. Using a hand-held compass with vertical sights (a primitive azimuth compass) he took regular magnetic bearings of every possible sunrise and sunset. This enabled him to compile a magnetic chart and, over time, to discover that magnetic variation was continually changing.

But even after Halley had published his *"A New and Correct CHART Shewing the VARIATIONS of the COMPASS in the WESTERN AND SOUTHERN OCEANS as observed in ye YEAR 1700 by Edm. Halley"* and *"Advertisement Necessary to be Observed in the Navigation Up and Down the Channel of England"* in 1701, many mariners still regarded compasses with suspicion.

The captain of the 70 gun *Lenox*, (one of the advance party in Shovell's fleet that had avoided the Scillies – see below), returned no less than 10 of his compasses to Chatham dockyard *"broke and in pieces"*, and his officers testified at the enquiry into the disaster that the others still on board were useless. The difficulty of holding a compass course when lit only by a candle, or in heavy weather, the proximity of metal, the need to "swing" a new compass properly, the inability to use a hand-held azimuth compass to check variation (because they could not determine latitude correctly); all were drawbacks that delayed acceptance.

Even today a magnetic compass can only be used effectively to navigate across or over oceans if in possession of ephemerides; or in conjunction with other navigating tools that also rely on advance prediction tables calculated from static observatory data.

The longitude prize fiasco.

The Admiralty's insistence that Newton's invention remained secret - based partly on the notion that as British sailors were the best navigators on the planet without needing the assistance of his instrument, so why let lesser nations in on the act? - was indirectly responsible for the Shovell disaster.

Shovell, Halley's commander-in-chief at the time, had ignored Halley's notification that the Isles of Scilly were 10 miles south of the current mapped location and that the magnetic variation had changed. Although Halley was spot on, Shovell had already made clear his very low opinion of academics posing as naval captains.

The Shovell disaster was responsible for the next Admiralty bright idea. In 1713 Queen Anne was persuaded to put her name to a prize fund offered by the Admiralty through a Parliamentary Act, *"providing a Publick Reward for such persons as shall discover the Longitude at Sea"*. Basically the sum of up to £20,000 (about £3 million today) for anyone who could manufacture an instrument that could be sailed from an English port to a Caribbean one, and there used to determine the longitude of said West Indian location.

The Admiralty was also indirectly responsible for the construction of yet another set of cyphers.

Christopher Wren sent Newton, chairman of the Longitude Board, a three line cypher which, when decoded appeared to be a claim or claims on the Queen Anne Longitude prize.

OZVCVAYINIXDNCVOCWEDCNMALNABECIRTEWNGRAMHHCCAW
ZEIYEINOIEBIVTXESCIOCPSDEDMNANHSEEPRPIWHDRAEHHXCIF
EZKAVEBIMOXRFCSLCEEDHWMGNNIVEOMREWWERRCSHEPCIP

The solution is simple enough. Read from the right in each line, transferring every third letter to a second list. The result relates to two (or three) inventions of Wren (and the late Robert Hooke) that, so Wren appeared to believe, entitled him to a share of the prize, miss-spelling notwithstanding. The second list merely contained the date and name of the applicant.

The three claims were related to a watch in a vacuum that Wren already knew was useless at sea; a telescope on gimbals that, when tested by Wren and Hooke had not worked either, and a method of checking a sailing vessel's distance run that could make no allowance for currents or drift. As far as is known, Newton did not bother to reply.

On a rare occasion that did merit a reply, Newton sent a waspish but apposite note to the British Admiralty Secretary Josiah Burchett who had unwisely suggested he consider an unfrocked clergyman cum watchmaker for a portion of the longitude prize. The fact that said ex reverend was also advocating that the Day of Judgement was at hand and that the influential Burchett had been responsible for placing that embargo on Newton's instrument did nothing to improve matters.

"And I have told you oftener than once that it is not to be found by Clock-work alone. Clockwork may be subservient to Astronomy but without Astronomy the longitude is not to be found. Exact instruments for keeping time can be usefull only for keeping the Longitude while you have it. If it be once lost it cannot be found again by such Instruments. Nothing but Astronomy is sufficient for this purpose. But if you are unwilling to meddle with Astronomy - the only right method and the method pointed at by the Act of Parliament - I am unwilling to meddle with any other method than the right one."

Newton was of course, absolutely correct on all points, including the fact that longitude could not be *"found by Clock-work alone"* for the reasons stated; and astronomy *would* be an essential ingredient. Newton's comments were made in 1721 but the longitude problem was not solved for another 55 years.

The Queen who's name was attached to the longitude prize regulations had died shortly after the ink had dried on the parliamentary act and Newton, who had headed the committee that had drafted those regulations died in 1727. During the 13 years that he had presided over funding applications for methods of discovering the longitude at sea, none had merited serious consideration. The infighting that then followed and eventually resulted in John Harrison's brilliantly constructed chronometer H4 winning *part,* but not all of the prize, is documented elsewhere, but a number of significant points merit inclusion here.

Newton and Halley (who was also party to the rule drafting) both knew that no one could ever win the full prize because Newton had already invented an essential marine octant (and the rules precluded anything already invented). They also knew that the longitude at sea puzzle could not be solved until the *latitude* at sea problem was solved; and that Newton's invention was the gem that would solve both at the same time.

The two also knew that there were two ways to solve this joint puzzle: either by the "lunar distance and marine octant method" or by the "watch and marine octant method". But the contestants themselves were not to know this for certain, although Nevil Mascelyne, being an astronomer and eventually to become the 5th Astronomer Royal, must have realised.

In 1732 one of those contestants was John Hadley, a vice president of the Royal Society. Hadley went to considerable trouble to set out his stall, persuading the Admiralty (who were responsible for funding the Queen Anne prize) to set up sea trials for a marine octant of his invention.

The Longitude Board co-opted the then current Astronomer Royal and vice-president of the Royal Society (Halley!) as an official observer to sea trials aboard the Admiralty vessel *Chatham* in the Thames estuary. The trials failed miserably; but looked good on paper after the data were manipulated. Two years later Hadley had the cheek to obtain a patent on *"An instrument or quadrant for taking at sea the altitude of the Sun etc.,....."*. Patent No 550 described Newton's instrument closely but was primitive in reality. The patent also suggested that the sea trials were a brilliant success.

However Hadley never received a penny from the board but is to this day credited by many as being the inventor of the marine octant.

http://trailblazing.royalsociety.org/

Because of the aforesaid shenanigans, the real breakthrough took until 1776, when Mascelyne succeeded in publishing the long-awaited lunar ephemeris that finally enabled mariners to determine position using the Moon as a heavenly clock.

By now John Harrison had just proved his marine chronometer fully seaworthy and suddenly Great Britain had acquired a virtual monopoly on the ability to navigate globally by two different methods, most importantly on the high seas. Use the easy method of chronometer and noon Sun height daily (in conjunction with a solar almanac), and the 3 or 4 hour-long mathematical task of computing position by the lunar distance method when conditions permitted - which incidentally allowed for *checking the accuracy* (rate of change) of the ship's chronometer.

Harrison had supplied the watch and Mascelyne the lunar distance data. Both methods complimenting each other, but upgraded commercial versions of Newton's octant, a marine sextant was essential to either.

It had taken the most intelligent species on Earth three millennia to progress from candle clocks to reliable chronometers, solar, stellar and lunar almanacs and self-correcting angle-measuring devices. Thus to discover how to retain the essential "home port time" whilst travelling across uncharted territory. Both solutions suddenly arriving within 2 years of each other; a classic example of waiting 3,000 years for a ride, only to suddenly receive two simultaneous offers.

Now navigators leaving London could set their chronometers to Greenwich *mean* time and, with luck, carry this time with them all the way round the planet (as James Cook did). Checking it by lunar tables and marine sextants (now no longer banned, but in many cases still requiring Admiralty permission to purchase) or Jovian moons on shore whenever the opportunity arose.

It is easy to forget that Newton's brilliant idea was responsible for solving four marine navigational conundrums, all linked to the problem of taking accurate angular readings from the deck of a ship. It was self-correcting, it could be used in almost any weather conditions, could take accurate angular readings of Sun or stars and, when laid on its side, could gauge the distance from land.

Portions of Chapters 3 and 4 were first published in *Astronomical Minds* in 2007, and later at a National Maritime/Royal Institute of Navigation conference and in *Navigation News*, the bi-monthly publication of RIN in 2010.

PART TWO

WHAT IS LEFT MUST BE THE TRUTH ?
THE PECTEN STRUCTURE HYPOTHESIS

How can each of those BTO cuckoos mentioned in Chapter 2, navigate from the UK to central Africa in directional fits and starts, and return the following spring, often by a completely different route? How can it be that these migratory navigational manoeuvres vary from one year to the next?

As has been indicated, it is impossible for long-distance avian migrants to get from unknown point A to unknown point B by the three compass method. So how does a cuckoo manage to fly (or walk even) all the way from Scotland to central Africa or wherever - and back again?

Surely there must be an entirely different explanation. As the late Sir Eric Eastwood, when discussing confusingly complex radar tracks of migrant birds in his book *Radar Ornithology* way back in 1967 wrote *"...the memory store and programme of the bird's innate navigation computer required to accomplish all these migratory manoeuvres has become bewilderingly complicated. One feels intuitively that a simplification of the (map and compass) theory must be possible somewhere"*

Part Two outlines one way in which Eastwood's "simplification" could be achieved. Not, of course, by the three compass method, *but by that innate phototactic escape mechanism which produced those original caged experimental claims* set out in Chapter 1. Like many simple solutions, this one appears exceedingly complex. Part Two also examines similarities between Isaac Newton's marine octant and the avian pecten structure.

Note. Much of the text and many of the diagrams in Chapters 6 and 7 and Appendixes 1-3, first appeared in print in Part One of the author's 1981 book *"Instinctive Navigation of Birds"* (now out of print).

Chapter Five

Organic Sun Visors

As anyone driving directly towards a bright sunset knows, clarity of vision can be badly impaired by solar glare. Time to clean the windscreen and bring the visor down. Walk eastwards at first light and you will either have to pull the rim of your hat down, use a hand as an eye-shade or risk tripping over something. Adaptive evolution has come up with some interesting measures to maintain clarity of vision under such testing circumstances.

Trilobites.
No one is sure exactly how trilobites became trilobites in the first place; they just suddenly appeared, fully functional fossil records embedded in Cambrian shale deposits (Figure 5.1).

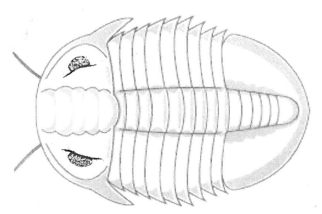

Figure 5.1 Trilobite outline.
Viewed from above. Legs not shown.

These already numerous species of trilobites were living in shallow seas about 540 million years ago, already possessing sophisticated visual systems made from glassy clear calcite crystals. These could respond to incoming rays of light via receptor cells, feeding a mosaic of tiny images; one per calcite lens, to the brain (Figure 5.2).

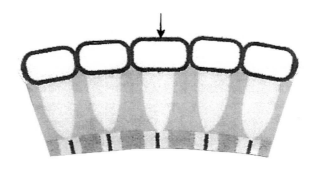

Figure 5.2 A section of the eye of a trilobite.
Rays of light are thought to pass through individual calcite lenses to crystalline cones and along nerve fibres to a primitive brain.

Unfortunately neither receptor cells nor brains can be converted into fossil material, so the degree of primitiveness is speculative. Probably trilobite eyes functioned on the same principle as those of modern arthropods and like these, some species possessed photo receptors that were far more efficient than others.

At least during this era there were no flying predators and those species of trilobites that chose to lumber slowly along the seafloor (rather than wallow actually in the mud) grazing on whatever trilobites grazed on, could do so in shallow water without fear of being dive-bombed.

The stuttering evolution of curved (wide-angled) clusters of these multi-faceted crystal photo-receptors, enabled trilobites to make the best of their constantly changing environment and each time a trilobite shed its armoured exoskeleton, another top row (dorsal layer) of calcite lenses was added. This increased the animal's light-registering powers and its chances of survival.

Nevertheless glare caused by an overhead Sun filtering through a shallow layer of water, would have severely hampered the ability of these dorsal rim detectors, and no doubt many trilobite species could only move freely at night.

Astonishingly, slap bang in the middle of the trilobite era, at least two species, by evolutionary good fortune, acquired "eyebrows" that protected eyes from a degree of overhead glare. One of these, *Erbenochile issoumourensis* or "big eye", also possessed 360 degree all round vision (Figure 5.3).

Figure 5.3 A close up of an eye of *Erbenochile issoumourensis*.
Note the overhanging rim and curved surface.

Finally, after continuing to evolve into myriad and sometimes wondrous forms - trident wielding weirdoes, species encased in armadillo-like armour or camouflaged as pincushions, and ranging in size from that of small dogs down to the size of gnats - the entire trilobite clan vanished again just as suddenly as they had arrived.

Insects.

Either trilobite mineral-based compound eyes were adopted and further adapted by insects or insects evolved their own independently. Certainly there are striking similarities.

Many modern-day species, including some ants, bees and butterflies, possess dorsal rim ultra-violet receptors, and almost certainly have done so for millions of years. For example, the honey bee's bank of dorsal rim facets appears to act as a very primitive short-range solar compass because they are receptive to UV light. This would enable a foraging bee to hold a course, once decided on, based on any specific polarisation pattern. On reaching an unfamiliar feeding area by this method, honey bees are known to fly backwards many times before settling on a reverse homeward course, presumably in order to check and then lock onto the opposing polarisation pattern. They have also been tricked into doing a loop-the-loop by laying a mirror beneath their flight path, reflecting the UV light upwards (Adrian Horridge, 2009).

Although insects have made good use of the ways in which their limited vision can be adapted to aid orientation, the underlying compound eye construction method ensured that vision has remained primitive, compared to the continuing evolution of the non compound lens-bearing eye. For example honey bee vision, at best permits pattern recognition in various shades of grey and a bee can no more view the complex outlines of a tree today, than its ancient ancestors could.

Birds.

The earliest known true birds began making airborne assents in the late Jurassic period some 140 million years ago, and were direct descendants of small feathered dinosaurs rather than of flying reptiles of the pterosaur type one sees pictured on the back of cereal packets.

However the puzzle as to exactly where and how the avian eye evolved is not so easily answered. This is because at least 30 different ways of registering light variation have come and gone and adapted and reinvented themselves during the time that life has existed on Earth.

But we live in hope that new forensic methods will one day discover exactly when a very large soft comb-like structure managed to put in an appearance directly in front of the optic nerve and projecting into the clear gel that fills the space between the flexible lens and the retina of said avian eyeball (Figure 5.4).

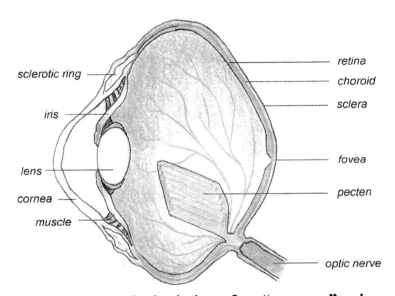

Figure 5.4 Schematic depiction of an "average" avian eye.
When bright light shines through the lens, the pecten structure casts a protective shadow on the delicate light receptive areas of the retina. But the retina is not protected from low-level glare.
Source: Wikipedia.

It has been suggested that the pecten structure can exert pressure on the lens and alter its focus, that it could assist colour registration or possibly be of use in any of more than a dozen other ways. But undoubtedly the pecten acts as a glare filter of a specialist kind that decreases glare intensity from all but the lowest areas of input. Sun visors at work again, not dorsal rim strips or peculiar trilobite "eyebrows" this time, but by the pecten structure filtering incoming light from higher angles.

This method of glare reduction would appear to create a new drawback. Single lens eyes are generally defenceless against incoming bright light from *low* angles, whereas bees eyes are composed of hundreds of facets most of which detect motion and bits of objects and can operate in bright sunlight without causing serious loss of vision.

If as is thought, the pecten structure can reduce glare coming from somewhere in the region of 30 degrees and higher, such a structure would conversely increase the blinding effect of low angle solar glare (H.B.Barlow and T.J.Ostwald, 1972). Equivalent to driving towards the rising Sun by peering *beneath* the car visor. Almost as blinding to a bird as to a mariner using a Newtonian style sextant when trying to view the image of low level Sun as mentioned in the Preface.

The following two chapters explain how the mere possession of a pair of eyes, each containing a pecten structure lacking the ability to filter low-level glare, can conceivably enable long-distance avian migration to be accomplished innately.

Chapter Six

Flying Long Distances Without a Compass
Courtesy the Pecten Structure

The pecten structure must have played such an important role in avian evolution that this has far outweighed those disadvantages mentioned in Chapter 5. Far more important than enhancing colour vision or aiding focussing or providing extra blood supply to the retina or even protecting the retina from bright sunlight.

Menotactic response to light.

One-eyed response to light is no recent discovery, and neither is the specific choice of one particular eye. More than a century has passed since the Swiss entomologist Felix Santschi identified this "preferred eye" behaviour in ants (Figure 6.1). He also discovered that ants could not home successfully in heavy overcast conditions.

Figure 6.1 Mirror experiment with a homing ant suggesting specific menotactic behaviour.

At points 1,2,3 and 4 sunlight was intercepted and the image was projected from the opposite side of the ant with the aid of a mirror.
Source: Derived from F.Santschi (1911).

Although the menotactic behaviour of homing ants has little to do with the directional reactions of birds when confronted with low-level blinding light, it does serve to remind us that one-eyed directional responses within the animal kingdom has long been noted.

Sections of large overnight roosting flocks of dawn-dispersing starlings sometimes appear to exhibit this one-eyed response when confronted with a glaringly bright sunrise. Attempting to head *directly* back to yesterday's favourite feeding areas into the sunrise seems temporarily impossible and a short-term diversion during which one eye is blinded and the other provided with clear vision, is forced on that particular group (Figure 6.2). A one-eyed bird would find itself as disadvantaged as a mariner lacking an effective low-level filter for his sextant, but the bird with one pecten filter on each side of its head can cope.

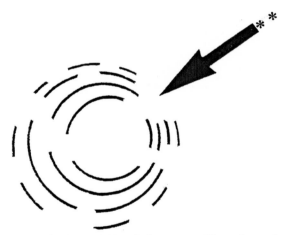

Figure 6.2   A typical starling "ring angel" radar plot of a roost dawn dispersal.
Showing successive waves of dispersal, and effect of wind in relation to the distance covered by different groups.
Source: Derived from E.Eastwood (1967) Gerrard (1981a).

Menotactic behaviour in birds is not confined to flight. Examples of stationary fish-hunting species such as herons being forced to tip the head to one side in sunny conditions is well documented. In one eye the reflected glare off the water surface is reduced as is the brightness of the solar rays. The fact that herons can still be observed behaving like this on sunny days when the water surface is rippled, suggests the Sun's dazzle still has to countered (Figure 6.3).

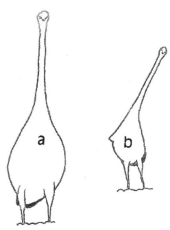

Figure 6.3   Head-on view of two hunting herons.
Drawing (**a**) was made on an overcast day,
drawing (**b**) on a sunny day.
In drawing (**b**) the bird is tilting its eye downwards towards the Sun's direction, and thus moving the pecten structure's position so as to avoid being dazzled.
Source: Derived from J.R.Krebs and B.Partridge (1973).

On the Welsh island of Bardsey, the lighthouse, which until adapted, sent regular pulses of useless light across the mountainside, acted as a death trap to nocturnal migrants each autumn.

Why were these thousands of migrants attracted to this lighthouse beam, and why did the fatalities decline when the light was prevented from illuminating a mountain? They were attracted to the light in the same way that moths are attracted to a candle or an illuminated white sheet above a moth trap. Unable to break free from the mesmerising "pull', many spiralled in and collided with the light's glass protective covering. Knocking themselves senseless or worse.

Why Bardsey in particular? The larger-scale attractions occurred on nights of low overcast, conditions unsuitable for southerly-flying nocturnal migrants to continue with any degree of enthusiasm.

Forced either to fly on into mist or cloud - a big "no-no" for birds - or descend, they were then initially attracted *directly* by the bright light, flashing slowly round and highlighting the mountainside at or just below eye level.

Then on nearing the light source, they were forced into a menotactic induced flight and blinded in one eye by the failure of the pecten structure to mask the glare from lower levels and seeing little if anything in the unblinded eye. Spiralling round and round, colliding with each other or the lighthouse and eventually dying, collapsing or escaping the trap (Figure 6.4).

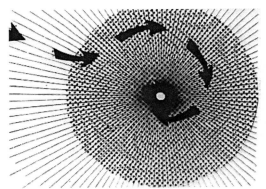

Figure 6.4   Attraction of low altitude flying migrants to lighthouse glare.
Indicating how forced menotactic response results in a
collision through maintaining the angle when the
overriding object of attraction is attainable.
Source: Gerrard (1981a).

On nights when low cloud was absent there were fewer fatalities but often just as many landings on the island. On nights of really bad weather most migrants were not aloft and the lighthouse was free to do the job it was designed for.

With rapidly increasing numbers of man-made structures now creating low-level artificial glare (lighthouses, oilfield gas flares, illuminated skyscrapers, airport ceilometers, laser beams and who knows what else?), the avian pecten structure is causing ever-increasing self-induced fatalities and the mercifully infrequent aircraft accident. Not too much damage caused by a large aircraft colliding with a swarm of confused moths on take-off but meeting a flock of disoriented migratory geese on a night with a low cloud base is an entirely different matter. *This problem cannot be circumvented by pretending the peculiar construction of the avian eye presents no danger to aircraft in such conditions.*

How the shadow-casting properties of the pecten structure can unwittingly "guide" any long-distance avian migrant from one unknown location to another must, by definition, be a very simple process for a bird. **First the diurnal migrant.**

The noon Sun signpost towards lower latitudes.
At all times of the year outwith the tropical zones, the position of the Sun at its highest (local noon) points towards the equator. The higher the latitude and the nearer to the respective winter solstice, the more obvious this is. If the Sun at its highest were to provoke a positive phototactic response in higher latitudes, a directional bias towards lower latitudes would be induced (Figure 6.5).

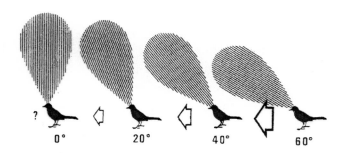

Figure 6.5   The relative influence of noon Sun angle at various latitudes.

Source: Gerrard (1981a).

At the times of the autumnal and vernal equinoxes, the Sun rises in the east and sets in the west in all parts of the world with the exception of either pole, in the vicinity of which there is a distinct lack of budding migrants anyway. But at other times of the year the sunrise and sunset directions vary in tandem predictably just as the daylight hours do. For example if the Sun rises 10 degrees to the south of east on a certain day, it will set that evening almost 10 degrees to the south of west (unless one is living in a tent in the shadow of Everest).

If at a certain latitude, the sunrise/sunset angles had been noted on or about the 23rd October, a similar event would occur again on the 20th February, i.e. a month after the (northern hemisphere) autumnal equinox and a month before the spring (vernal) equinox. The higher the latitude, the more these angles are going to change day by day (Figure 6.6).

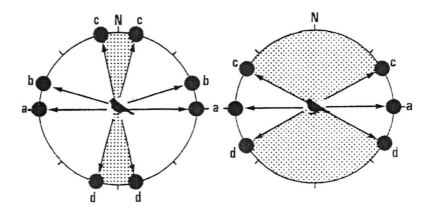

Figure 6.6   Sunrise and sunset directions at different latitudes.
Sunrise/sunset directions relative to an observer at 60 degrees N (left diagram) and 40 degrees N (right diagram).
a. Directions at either equinox. **b.** late April. **c.** Midsummer.
d. Midwinter. The Sun never rises or sets in directions covered by the shaded areas. Angles would be reversed at the dates indicated in the southern hemisphere.
Source: Gerrard (1981a).

So a potential migrant, in-flight travel bag packed and urgently feeling the need for a holiday, is about to set off early one morning attracted directionally by the Sun. If it spent all day flying directly in a mesmerised fashion towards that big bright orb in the sky it would have flown from east to south to west in a big U-shaped arc burning a great deal of fuel and achieving zilch in terms of mileage towards its winter quarters.

How then might the pecten structure straighten that useless U-shaped flight? Enforced menotactic (one-eyed) response to low-level solar glare, combined with direct attraction (to the Sun) at other times would seem to be the most likely answer. An explanation that immediately highlights another problem!

Flying from sunrise for 5 or 6 hours will result in good progress towards lower latitudes. However if a migrant is induced to fly all day, it will have to "switch" eyes in the mid-afternoon if much progress towards lower latitudes is to be accomplished (Figure 6.7).

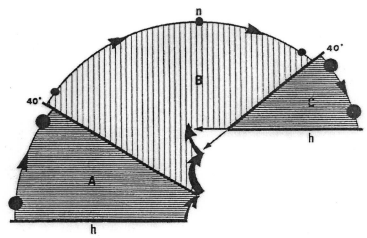

Figure 6.7 Theoretical day-long heading.
Theoretical day-long heading induced by a combination of morning glare **(A)**, followed by daytime direct attraction **(B)**, concluding with switched eye evening glare **(C)**. **h**…horizon. **n**…noon.
Source: Gerrard (1981a).

The vast majority of diurnal migratory flights commence at or near dawn and rarely last long enough to require an eye "switch" in order to maintain a longitudinal heading. On those exceptional occasions where a day-long flight is enforced because, for example, there is no acceptable place to land, the migrant/s may indeed respond with the wrong eye and find themselves making little useful progress.

On the other hand wave direction or wind speed and direction may influence the confused into continuing onwards (and automatically selecting the "correct" eye) rather than making that futile "U" turn.

Possible aids to maintaining a positive migratory direction.

Should any wind-borne object become attractive to a flying bird (or insect or even a pilot) this will eventually result in a change of heading to downwind. Cloud pattern attraction could thus persuade an otherwise confused avian migrant to fly downwind whenever there was no other more attractive object in sight.

Any specific cloud could normally be reached or overtaken, but if a string of clouds were the attraction, speedy over-the-ground (or sea) progress would be accomplished (Figure 6.8).

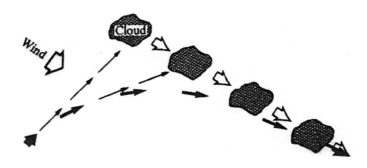

**Figure 6.8   A heading alteration to "downwind"
induced by attraction to clouds.**
Source: Gerrard (1981a).

However, if in overcast conditions, a flock of migrants were attracted to a distant *fixed* object in their line of sight, they could fly directly towards it, and on nearing this, could well be encouraged to select the next objective to the fore. This would produce a day-long flight in approximately a useful direction. But they could only fly directly towards it in a straight line if there was no appreciable crosswind.

The higher the crosswind component relative to the birds' flying speeds, the more likely the realisation that not much progress was being made in what had become unfavourable conditions and the more likely a halt would be induced.

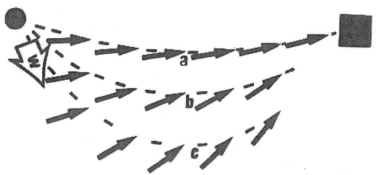

**Figure 6.9 An unwitting turn into wind when heading for a *fixed*
attraction during unfavourable crosswind conditions.**
a. fastest flyers. b. medium speed group. c. slowest flyers
Source: Gerrard (1981a).

The actual *headings* of each of the 3 groups illustrated in Figure 6.9 are
very similar. This would not be the case if the object of attraction was
unattainable: a star for example, as will be examined in the next chapter,
where it is proposed that nocturnal migrations can still be directionally
influenced by solar glare.

Chapter Seven

Flying Long Distances in the Dark Without a Compass Courtesy the Pecten Structure

The directional influence of the varying effects of solar glare on the pecten structure can obviously play no direct part in nocturnal migrations; excepting when encountering brightly lit man-made objects as already discussed. How then can long distance avian migrants travel anywhere directionally advantageous during the hours of darkness? How did those unfortunate individuals ever get, for example, as far as Bardsey Island in the first place?

The directional influence of low-level stars or star patterns.
Many migrants set off in the early morning, usually in flocks that provide some safety from aerial predators. Other species are reluctant to launch into unfamiliar territory in daylight and wait impatiently until oncoming night masks this daunting sector of the journey. Agitated individuals, often perched in trees, will align themselves more or less at right-angles to the glare of the setting Sun – *a menotactic response* which leaves one eye retaining clear vision (Chapter 6 Figure 6.1).

As darkness descends, they take off, joining thousands of others, calling excitedly to each other by way of encouragement; or fearful of losing contact. Whether they all head off towards lower latitudes or exactly the opposite, depends on whether the majority are aware of the terrain they have previously crossed, whether the high pressure system clear skies and temperature drop that triggers autumn migration is present, and whether they have been influenced by adults who may be familiar with the departure area from previous visits. Wrong-way flights of both juveniles (often) and adults (uncommonly) do occur.

Having taken off under clear skies in roughly the right direction –what next? Assuming most juvenile migrants have about as much intelligence as your average domestic chicken (possibly less), all they would have to do is to lock onto some attractive object to the fore – just as a chicken can be persuaded to follow a chalk line. Simply head for the brightest spot of light in the sky immediately in front of you and keep taking the tablets – a direct *phototactic escape response.* An innate response exhibited by all manner of animals.

Star following is not to be recommended if trying to hold a course whilst piloting a yacht, although it can provide temporary respite from constant compass watching. All natural heavenly objects appear to move across the sky nightly owing to the spin of the Earth, and following a bright point of light will result (at best) in a slow but constantly changing heading.

Locking onto a prominent star (or planet) that is at a maximum altitude of about 25 degrees (i.e., without having to tip ones head upwards at much of an angle) will not always result in a gradual westerly shift. In fact such stars appear to move at varying speeds and in a variety of directions depending on both latitude and the direction one is facing, as outlined in Figure 7.1. To avoid too much confusion, the directional influence that could be imparted when not facing roughly north or south is not shown, but is discussed in the next chapter when examining high latitude responses.

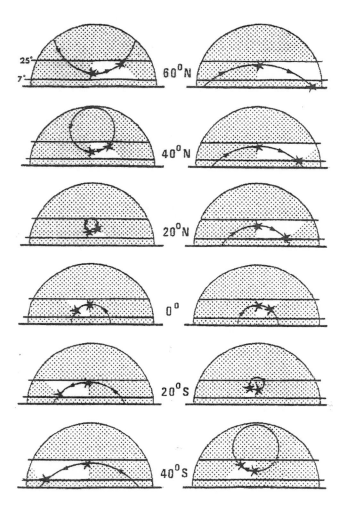

Figure 7.1 The directional movement of stars at various latitudes.
Left column - facing north, right column - facing south. The clear areas
indicate both direction and relative length of star movement
during any three hour period if viewing only those between
approximately 7 and 25 degrees of altitude.
Source: Gerrard (1981a).

This star-following explanation poses another question.

Does the Moon exert a directional influence on nocturnal migrants?
It is unlikely to trigger a menotactic one-sided response because it does not produce blinding glare.

Radar-based evidence covering sea areas where land was at best a long way off has indicated that migrants of various species were unlikely to have been influenced by the Moon. Then again, many nocturnal migratory movements take place in the absence of the Moon and those directions appear similar to those taken on moonlit nights, so lunar influence is still open to question. See Appendix 1 for further discussion.

The effects of crosswinds on nocturnal migrants.
Migrants heading for an attainable objective in crosswind conditions (day or night) will, with a degree of good fortune, reach it (Chapter 6 Figure 6.9). Star attraction will produce completely different arrival stop-over areas depending on the flying speeds of different species.

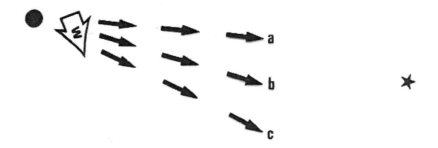

Figure 7.2 *Tracks* of groups with differing flying speeds.
These unwittingly shift across Earth's surface in different directions during crosswind conditions if all are attracted to similar unobtainable objects. **a**…fastest flyers. **c**…slowest flyers.
Source: Gerrard (1981a).

In crosswind conditions, some pioneer radar observers were persuaded that such migrants were *deliberately* compensating for this lateral displacement. However this notion should have been laid to rest way back in 1967 when the late Sir Eric Eastwood wrote a paper quashing that notion, entitled *"The interpretation of radar evidence relating to wind drift correction by bird migrants"* which was accepted by *Ibis* but never published.

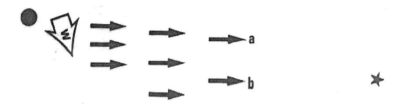

**Figure 7.3 In conditions as outlined in Figure 7.2,
the actual *headings* of all the birds are the same.**
Source: Gerrard (1981a).

Overland nocturnal migration.
Many may select distant landmarks as mesmeric aiming points; especially during clear settled weather. A number of radar studies have indicated this preference for ground-based objectives, confirmed by the tracks as was illustrated in Chapter 6. The actual objectives are not so readily revealed and have often been suspected as being man-made. The large numbers of different migrant species recorded in New York's Central Park every autumn (fall) must surely be because of the vast array of local night-time illumination surrounding the park.

Nocturnal over-sea flights that continue after dawn.
Living in the Hebrides provides a unique perspective on avian migration for a number of reasons – an unusually mild climate given the northerly latitude – unreliable wind speeds and directions – a fly-way for species breeding as far to the north-west as Greenland and to the north-east as northern Russia. Plus the opportunity to link the arrival and departure of these long-distance flyers with the prevailing weather patterns.

M.T.Myers (1964), through his radar studies in the Shetlands, discovered that in the absence of an easterly component in the wind direction, virtually no autumnal nocturnal migrations were observed to the *east* of the islands. However when the wind was from the east (from NE through to SE), considerable migration was often observed, much of which appeared to originate in Scandinavia.

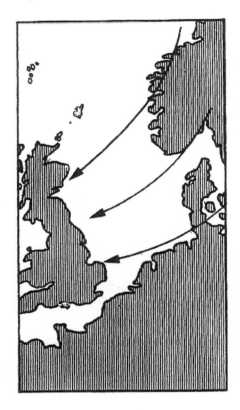

Figure 7.4 Typical "settled" high pressure weather conditions in north-western european waters.
Clear skies and clear nights initiate autumnal migration. There is often a change of wind direction from NE to E in lower latitudes.
Source: Gerrard (1981a).

Although many autumnal migrants would have initially been *heading* variously south, once over the sea in a gradually developing crosswind, such migrants would find themselves - aided by west moving stars - tracking westwards towards the British Isles. The stronger the easterly wind component, the further north autumn migrants are likely to arrive.

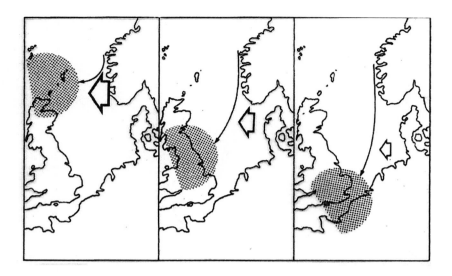

Figure 7.5 The stronger the easterly wind component, the further north autumnal migrants are likely to arrive in the British Isles.
Source: Gerrard (1981a).

During less than perfect weather conditions, on numerous occasions in late autumn particularly, flocks of migrants from the north-east have been registered on radar still far out to sea and still moving towards Scotland on a south-westerly track as dawn approaches. Quite suddenly they begin to gain altitude - probably having realised the "ground" beneath them is not suitable for landing on!

Some flocks then change course quite abruptly. Although the Scottish coast may just have been visible, such course changes are potentially disastrous but most likely were caused by the sudden appearance of sunrise and its attendant enforced menotactic response. If this enforced change is to the south-west all may well survive, but probably not so if a bad decision "eye-wise" is made that results in a backtracking to the north-east.

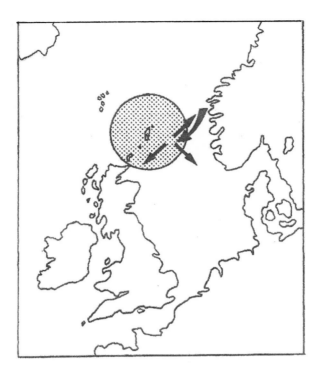

Figure 7.6 An example of dawn reorientation observed from a radar site in the Shetlands.
Curved arrow represents nocturnal tracks;
straight arrows, the subsequent dawn tracks.
Source: Derived from M.T.Myers (1964) Gerrard (1981a).

This "dawn ascent" behaviour was also noted by Eastwood and David Lack from radar sites on the east coast of England. More recently, with the discovery of oil and gas all across the North Sea, the picture, such as it is, has become very confusing. This is because bright beacons of light from rigs and gas flares have attracted tired and confused avian migrants experiencing poor flying conditions, luring millions to their deaths by gassing or incineration. Figures are naturally extremely hard to come by.

For species that can land on the sea (including numerous wader species that are normally reluctant to get more than their feet wet), sudden meteorological changes during migratory flights over water can be handled with less risk.

Encountering cloud whilst migration is in progress.

Birds, like light aircraft lacking all-weather instrumentation, tend to avoid flying in cloud. Indeed there are few if any records of such behaviour.

There are, however, numerous examples of altered responses on being confronted with a lowering cloud base. Either descend and fly beneath the base or ascend and fly above; and if neither is possible, make a controlled landing, backtrack or generally mill around in apparent confusion. "Apparent confusion" because species have reacted differently and nearly all those involved subsequently sorted themselves out. It was Eastwood's radar recordings of these "every which way" antics that provoked his comments mentioned in the introduction to Part Two.

PART THREE

AVIAN MIGRATORY EVOLUTION

Without the aid of pecten structures, birds could not have successfully countered climate changes by migrating across *unknown* terrain. A juvenile cuckoo lacking instinctive directional responses to light could never have flown from unknown point A to unknowable point B.

Even so, many avian species can be expected to respond in different instinctive directional ways. An eagle, buzzard or sparrow-hawk with wide bifocal vision (eyes facing forward), a pigeon or shearwater with narrow bifocal vision (eyes on each side of the head) or a woodcock with almost all round vision (eyes almost popping out of their sockets on each side of the head) could each be expected to respond differently to low-level glare.

Then again a hawk might simply follow its migrating prey. A pigeon could be expected to avoid a sea crossing, whereas a shearwater would behave in an opposite manner. A cuckoo, harried by smaller migrants might be forced into flight, day or night and those self-same small migrants might be reluctant to expose themselves to hawks if flying any distance in daylight.

Just to add to the mix, one part of the retina (of both birds and *Homo sapiens*), has a denser concentration of receptor cells (the fovea) which, when focussing directly on an object, enables us to perceive sharper images.

Many avian species only have a single fovea but some have two, the second being located in a different part of the eye – they are thus bifovial. These species are exceptional judges of speed and distance in order to keep track of moving objects – hawks and pigeons, terns and swallows for example. A few terns and swallows even have three, which presumably is why swallows can catch millions of those ubiquitous Scottish flying midges when we only notice them after being bitten by one - or fifty.

Further questions and possible explanations can be found in Appendix 1

Chapter Eight

Maps and Geolocators
An Oceanic Avian Migratory Route Revealed

It was impossible to produce an *accurate* large area map of a globe (without a lot of curved gaps) on a flat surface. Compromises had to be made; and in 1569 Gerardus Mercator, whilst still of the opinion that Earth was at the very centre of the Universe, produced the first chart suitable for nautical use by drawing all the longitudes and latitudes as straight lines. Such charts did not display the shortest (great circle) route when crossing longitudes diagonally.

The Mercator projection chart was, until the 18th century, popular with mariners because a course could be plotted from A to B simply by drawing a straight line and then followed roughly, very roughly, by holding a single fixed magnetic compass bearing throughout the voyage.

Confusingly, such projections show Greenland to be about four times bigger than Australia when it is actually only one third of the size! So although direction was easier to plot, distance travelled required a good head for figures in order to convert the flat chart measurements into true distances covered across a curved surface.

When the author first published avian migration route maps in the 1970's, the Mercator system was chosen because anywhere along any of the *straight* horizontal lines of latitude on the chart (or along any one wished to add), sunset and sunrise directions were identical for any given date - sunrise in SE, sunset in SW for example.

It was an ideal tool to plot sunrise/sunset "lift off at right-angle" directions *if* avian long distance migrants were being influenced by the inability of the pecten structure to prevent low-level glare. If the known routes roughly matched these departure headings, the hypothesis *could* be valid, and this would confirm the birds were *not* taking the shortest, most fuel-efficient route. If this was so, *further range expansion could be predicted.*

One example relating to a species familiar with oceanic environments is set out below.

The sooty shearwater *(Puffinus griseus)* is one of the most abundant species of seabirds on our planet - 22 million or so. Those breeding in New Zealand lay their single eggs in burrows in late November, and depart (after first refuelling), in a somewhat leisurely fashion for their wintering grounds in the summertime of the north Pacific. There they moult before returning to New Zealand at a considerably faster rate, arriving from late September onwards.

With a figure-of-eight migration route that has been reasonably well documented, it had long been realised (60 years at least) that this route was underpinned by the seasonal wind systems of the Pacific Ocean. Remove the wind component and the assumed tracks would become more or less upright – north-south headings.

Then tracking systems were developed and geolocators that recorded pressure and temperature were fitted to considerable numbers of breeding-age sooty shearwaters. The results indicated that the figure-of-eight migratory route was not as straightforward as first thought. Almost without exception the birds commenced their autumn migration by heading east or even south-east; why?

http://www.pnas.org/content/103/34/12799.full.pdf

When the seasonal wind systems are examined it seems that the birds were not initially actually migrating, but merely dispersing *prior* to migration. They were flying *downwind* through and towards wonderful feeding grounds nearer the Chilean coast. As soon as the replenishment of energy reserves had been completed, migration could commence. The earlier (not so far east) this initial fattening process had been completed, the sooner the migration began and the further to the west (up in the north Pacific) it ended (Figure 8.1). Wandering within either of the two suitable wintering areas then commenced.

The return journeys back to the restricted area of the breeding grounds again appears to have started with a lateral energy reserve build up, prior to a full-on southerly migration. This was then accomplished by a reliance on the seasonal wind changes and more competitive urgency to reach the breeding grounds. Less wandering equates to less wind induced drift. The speed and timing of each individual journey is the key to the overall route variability year on year and imprint attachment to breeding sites completes the picture. An over-simplification lacking detailed evidence to date.

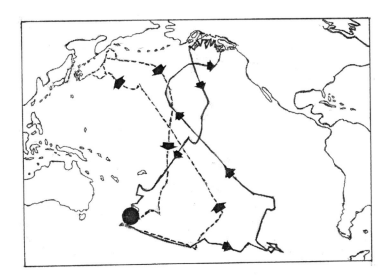

Figure 8.1 Tracks of two sooty shearwaters from New Zealand to the northern Pacific and back.
Selected by the authors from the 19 tracks to illustrate the different northerly routes taken, dependant to some extent on the original longitude at which migration commenced.
Source: Derived from S.A.Shaffer et al.(2006).

Note the similarity of the tracks in Figure 8.1 with lateral pre-breeding movements of various BTO GPS satellite-linked cuckoos discussed in Chapter 2 and illustrated on the BTO web site.

http://www.bto.org/science/migration/tracking-studies/cuckoo-tracking

The confusing migratory behaviour of the westland petrel *(Procellaria westlandica)* (T.J.Landers 2011) indicates that these birds simply fly downwind towards southern Chile as the sooty shearwaters are arriving, feed all summer in the Peruvian north-flowing coastal current, and fly downwind again back to New Zealand just before the sooty shearwaters are leaving as the winds switch direction once more. A balloonist could do this and the two-way "migration" seems to mimic the sooty shearwaters' post and pre-breeding lateral feeding movements.

A similar two-way "migration" linked to seasonal changes rather than driven by the pecten directional influence, is exhibited by the African river martin *(pseudochelidon eurystomina)*, a species that breeds far up the Congo river in the dry season and, when flooded out simply flies downriver to winter along the Atlantic coast.

The migratory tracks within the Atlantic Ocean of the great shearwater *(Puffinus gravis)* exhibit similar flight behaviour to that of the sooty shearwater in the Pacific. However the apparent pin-point arrival at their breeding quarters on the small isolated Tristan da Cuhna group of islands requires a further comment, and can also be likened to the behaviour of many colonial nesting sea birds.

Having spent time on and off observing the homing behaviour of Cory's shearwaters *(Calonectris diomedea)* to the small island of Salvagem Grande (part way between Tenerife and Madeira) from the deck of a yacht, the home-finding method is in essence remarkably simple. Apart from being well aware of the wind/wave directions, these beautiful and efficient flying machines quarter the ocean in their day-long hunt for food. Much of this time is spent alone but nearly always within sight of another individual. Nearing sunset everyone heads for home, always still within sight of another. Those who had fished within sight of home from the outset, high-tail it back. Those next nearest follow them and so on. The last in had been far out of sight of the island initially, but can home by following the visual trail.

Chapter Nine

Predicting Migratory Trends
Courtesy the Pecten Structure

The probability of rapid evolutionary changes in long-distance passerine migrants; the wheatear species.
Some long-distance migratory routes of species that will, unlike those discussed in Chapter 8, die very rapidly if they have the misfortune to land on water, have most certainly not been in use for thousands of years. The last ice age would have had a profound effect, slowly reducing the distance between breeding and wintering locations from many thousands of kilometres until both ends met and the survivors became sedentary breeders. The entire process could then be slowly reversed as the ice receded.

However each time a passerine (songbird) migratory species is subjected to a slow compression of range and then, thousands of years later, a slow re-expansion, latent evolutionary traits could be expected to reveal themselves at any stage. For example; adaptation of a permanent sedentary life style could give the less agile individuals at the height of the ice age a slight selective advantage. This might force the more adventurous or less aggressive individuals into being the ones seeking fresh breeding areas the moment the climate began warming again.

Back in the 1970's, unravelling the complex origin of many current migratory species, using the pecten structure directional influence as a tool, seemed just possible. In those days DNA was yet to be used to untangle ancestry; the fitting of tracking devices was confined to large mammals, and GPS, let alone geolocators were tracking tools still only dreamed of.

DNA testing has since provided genetic links that suggest many modern avian species are not modern at all, and no longer can it be assumed that the recent Pleistocene series of drastic climate fluctuations could throw up a host of new migrant species. This unfortunately means that at present it is not, after all, possible to look backwards with any confidence in order to better peer into the future.

The Songbird 30,000 kilometre two-way migration world distance record holder. The common wheatear *(Oenanthe o oenanthe)*.
Geolocators were again used, but fitted as a lightweight back-pack, omitting temperature and pressure sensors. Thirty adult common wheatears were caught at the end of the season on their Alaskan breeding grounds in June 2009, and three were re-trapped in the same localities almost a year later. When the data were downloaded, *some* of the details of the incredible long-suspected migration route to and from north-east Africa were revealed.

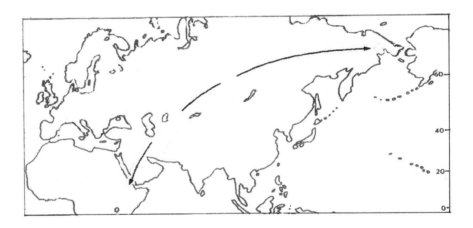

Figure 9.1 The common wheatear.
Map of the two-way migratory tracks between breeding grounds in western Alaska and wintering areas in north-east Africa.
The curving arrow represents the expected *average* two-way track of migrants under the influence of the pecten linked "right-angles to sunset" method of "navigation".
G.P.Dement'ev at al. (1954), provided the times at which migrants were observed at different latitudes in order to compute the anticipated sunset angles.
Source: Gerrard, (1981a.)

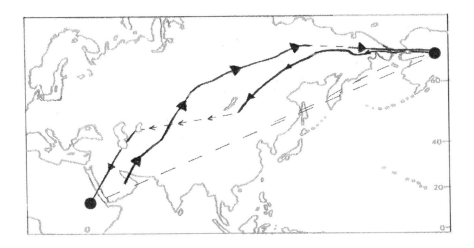

**Figure 9.2 The data obtained from one of the three
geolocator-fitted common wheatears.**
Compare the similarity of the route to the curved two-way arrow in
Figure 9.1. The 2 broken arrow sections = absence of data.
Source: Derived from H.Schmaljohann, et al.(2012).

The data in Figure 9.2 represents the recorded track of Bird C from the
Alaskan breeding grounds to north-east Africa in the autumn of 2009
and the spring return of 2010. The 2 broken lines where no useful data
were recorded was caused by the proximity of the vernal and spring
equinoctial periods when gelolocator data become totally unreliable.

The routes of Bird C were selected because by far the most firm data
had been collected. The autumn tracks of the other 2 birds (A and B)
matched the curve in Figure 9.1 more closely but the spring data were
very sparse.

http://dx.doi.org/10.1016/j.anbehav.2012.06.018   Summary only.

The straight dotted line between the breeding grounds and the wintering
area represents the course that would have been taken if using rhumb
line navigation; the dotted upward curve towards the north-east, if using
a constant magnetic heading. The great circle route is not shown but the

shortest distance between the 2 locations is via the North Pole. The authors compared every known *compass* course (6) believed to be available to avian migrants and the only one that more or less fitted the routes flown was a magnetoclinic course with a constant specific angle of inclination.

http://www.utm.utoronto.ca/~w3gibo/Models/Navigation%20models/magnetoclinic_hypothesis.htm

The Alaskan wheatear example is of special interest for several reasons, not least of which is that the author had published a map predicting the two-way route revealed by this recent geolocator data back in 1981 (Figure 9.1). This prediction had been based on the pecten influenced "right angles to sunset followed by nocturnal star following" hypothesis. The two-way arrow represents an average "out and back" route (hence the arrows at either end) because the two long journeys were known (from observer records) to be undertaken at different speeds and at different times relative to the equinox – differing sunset angles.

The magnetoclinic course with a constant specific angle of inclination has never been a method seriously considered for long-distance avian migrants; and one that does not fit the next case.

The Greenland/Iceland common wheatear *(Oenanthe o leucorrhoa)* – another record holder – this time for migrating back and forth across the north Atlantic.
K.Williamson (1953) had shown that the only common wheatears capable of migrating to and from Greenland/Iceland to Africa are those possessing better than average power-weight ratios, even with the assistance of favourable Atlantic wind systems.

As just illustrated, *nocturnal* migrants like the common wheatear leave their high latitude breeding grounds about a month *before* the autumn equinox and about a month *after* the spring equinox when departing the wintering grounds.

However the Greenland/Iceland wheatear somehow manages to winter to the *east* of its breeding grounds, not to the *west*. How can it do this influenced by the *same* pecten guidance system, and why not migrate down through the eastern seaboard of north America and winter there (Figure 9.3 C)? Surely a far less hazardous route for a bird unable to land on water, than trying to fly across an ocean and then on down to west Africa?

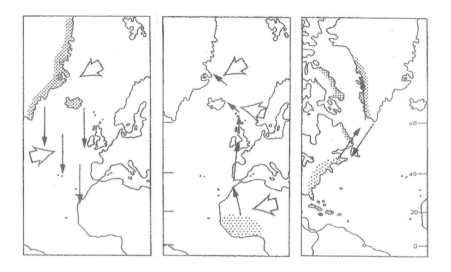

Figure 9.3 A, B and C.
A and **B**. Effect of wind on migratory directions.
C. Late departure temporary winter quarters.
Source: Gerrard (1981a).

Once the details of the route and the birds timing and responses are examined, the explanation for this wrong-way behaviour becomes apparent. They have no option, just as the Alaskan-breeding common wheatear had no option and could not fly to Indo-China, but had to make that enormous migratory two-way 30,000 km. plus journey.

In this instance the author had published three maps of the Greenland/Iceland birds in 1981 predicting the two-way migratory route (Figure 9.3), but had ignored the west Greenland and Baffin Island breeders through lack of any firm data.

Not until recently had the Alaskan team of researchers, again using geolocator equipment, managed to track a single bird from its breeding site on Baffin Island to west Africa and back. A second truly remarkable undertaking. This led to the claim by F.Bairlein, et al. (2012), that those joint results provided the first evidence of a migratory songbird being capable of linking African ecosystems of the Old World with Arctic regions of the New World.

Because the NW Atlantic birds commence their autumn migration about a month later than those in Alaska, the sunset angle has changed dramatically. This means they would commence their nocturnal flights *heading* more or less to the south of east, twisting westerly during the night by the low-level star shift (Chapter 7 Figure 7.1).

This twist back towards north America would be countered by the westerly autumnal western Atlantic prevailing winds, leaving the birds still to the south of the departure point at dawn with nowhere to land. Sunrise to the east – still head south and so on (Chapter 6 Figure 6.7) until land is sighted – which could be either the Faroe islands, Scotland, Ireland or even Norway; if they were fortunate.

Greenland departing birds would tend to arrive in western Scotland or Ireland and Icelandic birds in the Faroe Islands, eastern Scotland or Norway. Those departing from Baffin Island would probably either island hop across the Davis Straits to Greenland or even down to Newfoundland prior to initiating a full-on migration, and then across. Juvenile very late departures from Greenland or northern Canada would probably find themselves migrating down the eastern seaboard of northern America (Figure 9.3 C).

The author's own autumn ringing records of large Greenland/Iceland birds provided some substance to this explanation and entirely backs Williamson's findings. Certainly their wings are much longer and more pointed and even after a long sea crossing they are much heavier, sometimes almost double the weight of local migratory common wheatears, as well as being darker in colour.

So because this sub-species possess a strong imprint attachment to coastal areas; are powerful but *slow* flyers (something that enables the coastal and oceanic prevailing winds to exert a strong influenceon the migratory tracks), they can join, or even overfly the other branch of common wheatears from western Europe, to their African wintering grounds.

Thus it would seem, have the two long-distance migratory groups of common wheatears managed, entirely through the auspices of the pecten structure and basic low-level phototactic star attraction, to populate virtually all the northern ranges of our planet and at the same time avoid competing for winter space in and through the mid American bottleneck.

These examples underline the way in which long-distance migrant passerines could adapt to changing conditions far more rapidly than sedentary or partially migrant species could. *But in these cases there is no evidence of evolution actually unfolding whilst we observe.*

Chapter Ten

Unfolding Migratory Evolution ?

The chiffchaff/ willow warbler on-going evolution.
Close examination of the vast *Phylloscopus* clan of some 30 species suggests several intrusions into the Palearctic region from both Africa and Asia. But two of the long-distance migrant species are currently exhibiting signs of a widening gap between what surely was only recently one single species.

The chiffchaff *(P. collibita)* current breeding and wintering distribution, includes various isolated pockets of sedentary or semi-sedentary sub-species that appear to have evolved through lateral spreading along the lower latitudes.

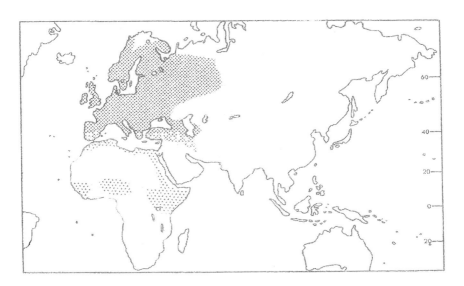

Figure 10.1 Chiffchaff northerly breeding and southerly wintering distribution.
Source: Gerrard (1981a).

Speculation. The northerly population overflow push has produced a short-distance migrant group in western Europe with a higher level spread, that has in turn produced a recognisable colour cline (darker to the west) and at least two sub-species. Another example of how avian migration could be spontaneously provoked whenever the conditions are suitable.

This has been promoted by an easterly spread of wintering grounds and consequent splitting of the migratory pool. This has blocked any real size cline development, because distance between breeding and wintering areas has not increased. All these groups have primary wing emargination on feathers 3 to 6 (Figure 10.3 A).

The willow warbler *(P. trochilus)* (Figure 10.2) could be considered as a distinct species or as a sub-species of the most recent chiffchaff expansion. The colour and wing clines and migration pattern about to be detailed, fit either alternative equally well.

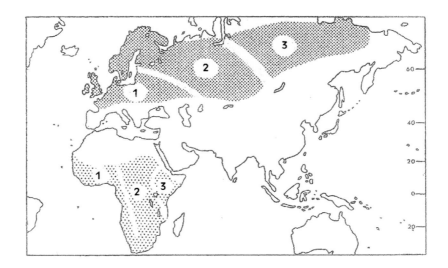

Figure 10.2 Willow warbler breeding and wintering areas.
1. *P.t.trochilus.* **2.** *P.t.acredula.* **3.** *P.t.yakutensis.*
Source: Gerrard (1981a).

The only morphological differences appear to be that the chiffchaff possesses emargination on the 6th primary whereas on the willow warbler this is absent (Figure 10.3 B) and willow warblers undergo a second full moult each year whereas chiffchaffs have one full moult plus a winter body moult. However some chiffchaffs also have a second moult of tail feathers and there have been reported incidences of interbreeding, so it would seem the two species are still in the course of full separation. Different songs? But so have separated groups of chaffinches.

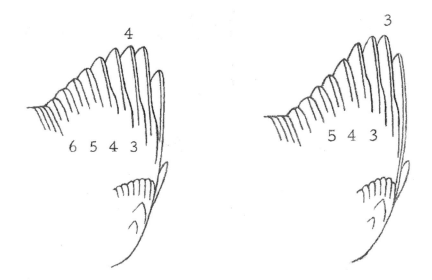

Figure 10.3 A and B. Wing profiles of chiffchaff (left) and willow warbler
Chiffchaff primary emargination on feathers 3 to 6.
Longest primary 4.
Willow warbler primary emargination on feathers 3 to 5.
Longest primary 3.
Source: L.Svensson (1970) redrawn.

Breeding ranges overlap considerably in areas where the chiffchaff is also a migrant, but the wintering areas of willow warblers are generally further south in Africa. There is a distinct size cline that can be related directly to migration distance. The final (?) push at present being undertaken is less obviously up into higher latitudes for two reasons.

Further extension is blocked by the discontinuity of the tree line in far northeast Siberia where the migratory distances back into southern Africa are as enormous as those of the more powerful and much larger common wheatear (Chapter 9).

The colour cline of the willow warbler is not as marked as in the chiffchaff and amongst the far eastern highest latitude breeding groups *(P.t.acredala* and *P.t.yakutensis)* these seem to be undergoing considerable upheaval (Figure 10.2).

The standard colour cline (lighter in the east) is being exhibited by the majority, but considerable dimorphism (differences between sexes) exists in certain areas. This is probably being caused by the restricted wintering area and could only sort itself out if alternative winter quarters could be established by *(P.t.yakutensis),* possibly in Indochina?

The sequence "allopatric first – sympatric follows" (allopatric; the evolution of a new species in different regions *first* and sympatric; one species splitting into two in the same region with no physical separation *second*) can be used to suggest the following sequence of events:-

During the last ice-age period that had denied the original willow warbler/chiffchaff ancestor much of its previous Palearctic range, only isolated pockets had survived. With the general range expansion into higher latitudes that followed the receding ice, at least one species split occurred and those with longer wings, less emargination and that double moult were equipped to occupy the distant breeding areas and overfly their more agile (more emargination), but less powerful counterparts.

So two sub-species could have emerged from the same gene pool, *before* the current chiffchaff and the willow warbler new breeding ranges began to develop, wherever the chiffchaff could reach them. Not in the far high east but readily in the west with its much shorter migratory routes.

On-going studies.

Migratory-based evolution tied to the pecten structure directional influence (or any other non "3 C" driving force) of avian species *generally*, is outwith the scope of this book. Research into several key groups has occupied various members of the Scottish Research Group for nigh on half a century.

These include the evolutionary implications of juvenile passerine lateral dispersal in general and the yellow-browed warbler *(Philloscopus inornatus)* in particular, as illustrated in Figure 10.4. Spread too far east and the species range could be extended to the east of the Himalayas and too far west and into west Africa.

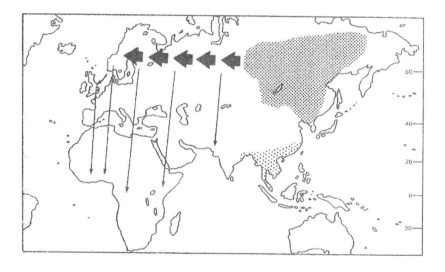

Figure 10.4 Yellow-browed Warbler breeding and wintering ranges.

Westerly juvenile dispersal and possible range expansion
Source: Gerrard (1981a).

Innate forced range expansion of Palearctic nocturnal longer-distance migrants as illustrated in Figure 10.5, is another on-going project.

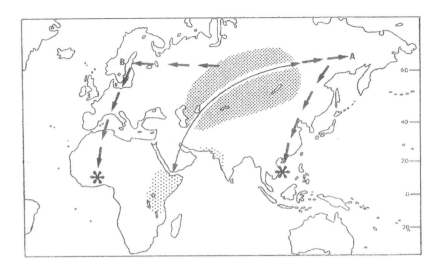

Figure 10.5 Instinctive forced range expansion of Palearctic nocturnal migrants when under population pressure.
A. Pressured breeding range expansion.
B. Pressured juvenile dispersal.
Source: Gerrard (1981a).

Another study relates to the starling range expansion after introduction into North America in comparison with that in Europe.

Ownership of twin pecten structures has provided countless thousands of different, constantly evolving avian species and sub-species with the innate ability to adapt their migratory routes from day one of their existence. A time, incidentally when there already existed seasonal areas at *different* latitudes suitable for birds to breed and winter.

DNA and survival of fittest for purpose is so wonderful a design that it is difficult to imagine all this occurring without the Great Creator being out there somewhere.

TAILPIECE – FULL CIRCLE ?

Changing horses in mid-stream

The Emlen cage pitfall trap.
Through the courtesy of GPS and geolocator tracking systems, it has recently dawned on an increasingly knowledgeable public, that long-distance migrants do not mysteriously spring up and fly, for example from northern Europe straight to Africa and back each year. Their routes appear to wander, as clearly proved by the BTO cuckoo tracking scheme and those cases mentioned in Chapters 8 and 9.

The time-served "3 C's" navigational methods uncovered courtesy of the Emlen cage are no longer in the running. A re-think is required; not via mass retractions, return of funding and the adornment of hair shirts, but by reappraisal of the original evidence.

A good example of a determined effort to avoid the former and embrace the latter is set down in the following paper :-

"A new approach to evaluate multimodal orientation behavior of migratory passerine birds recorded in circular orientation cages: A Ozarowska et al.,JEB.,216,4038-4046, 2013".

http://jeb.biologists.org/content/216/21/4038.full

The authors' are of the opinion that the basic Emlen cage methodology was a wee bit flawed. The manner in which all those scratches from all those individual inmates have been interpreted up to now, was *sometimes* in error because no allowance had been made for those birds that were confused over which of two different (multimodal) onward paths to take.

Although Ozarowska and three of the co-authors are Polish and another a Bulgarian, the fifth is Susanna Akesson of Lund University, Sweden and author and co-author of at least 12 papers based on data obtained by

Emlen cage techniques (Chapter 1 and Appendix 5), many of which are listed in the reference section of the paper.

This is the published summary with key points underlined.

Circular orientation cages have been used for several decades to record the migratory orientation of passerine migrants, and have been central to the investigation of the functional characteristics of the biological compasses used for orientation. The use of these cages offers unique possibilities to study the migratory behaviour of songbirds, but suffers from statistical limitations in evaluating the directions of the activity recorded in the cages. The migratory activity has been reported to vary, including complex multimodal orientation of migratory passerines tested in orientation cages irrespective of species studied. The currently applied circular statistical methods fail to describe orientation responses differing from unimodal and axial distributions. We propose for the first time a modelling procedure enabling the analysis of multimodal distributions at either an individual or a group level. In this paper we compare the results of conventional methods and the recommended modelling approach. Migratory routes may be more complex than a simple migratory direction, and multimodal behaviour in migratory species at the individual and population levels can be advantageous. Individuals may select the expected migratory direction, but may also return to safer sites en route, i.e. sites already known, which provide food and/or shelter in reverse directions. In individual birds, several directions may be expressed in the same test hour. At the species level, multimodal orientation may give an opportunity to expand the range or may refer to differential migration route preferences in different populations of birds. A conflicting experimental situation may also result in a different preferential orientation. In this paper we suggest a statistical solution to deal with these types of variations in orientation preference.

This new solution is illustrated via responses of 4 different warbler species, but one, the willow warbler *(Philloscopus trocilus)* is selected here because it's migratory route has already been discussed in some detail in Chapter 10, Figure10.2 etc.

Back in the autumn of 2001 representatives of the 4 different long-distance migrant species had been tested in Emlen cages in Bulgaria and the results had been computed in the normal manner, which suggested the willow warblers had on average attempted to head more or less SW – 229 degrees plus or minus 67.3 degrees to be a little more precise (Figure T.1).

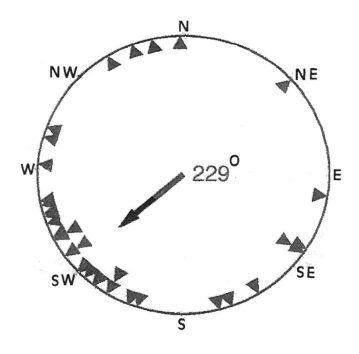

Figure T.1   Conventional procedure.
Willow warbler directional records. Mean direction 229 degrees with standard deviation (s.d.) of 67.3 degrees.
Source: Derived from A.Ozarowska et al., (2013).

But now a *second* method of analysing these results was applied. As the details of this alternative method occupies several pages of complex argument, it is easier just to outline the conclusion, one that resulted in a very different directional picture.

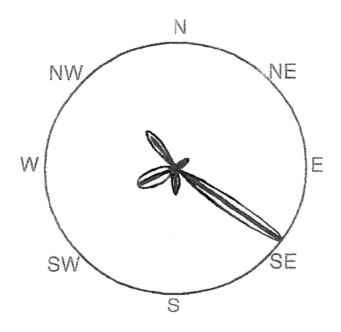

Figure T.2   Modelling procedure.
Willow warbler directional records.
Five alternative directions (in degrees).
57.1/s.d. 21.6.   **124.6/s.d 5.**   177.7/s.d 14.   244.6/s.d. 19.5.
322.2/s.d.8.3.
Source: Derived from A.Ozarowska et al., (2013).

Given the site of capture (44.00 N., 26.26 E.) and the option of resting up prior to one final dash for the African wintering grounds, the possibility that the willow warblers were unsure as to whether to depart towards the SW or SE, with the majority favouring the latter makes sound sense (if one is a willow warbler). OK so far, but the question raised is this. Figure T.1 demonstrates a strong pull to the SW, and thousands of circular diagrams based on Emlen cage directional scratches have done likewise using similar statistical analysis. Now the *same* data is reprocessed (Figure T.2), and now demonstrates a stronger pull to the SE.

They cannot both be right, and if this new method is the right one, much of the old material from possibly *thousands* of experiments must be flawed, as the author of this book has been claiming for the past 40 years.

The GPS pitfall trap.
Because of all this newly acquired GPS data on the tracks of long-distance migratory birds, several experts have now adopted a different approach to that chosen by the authors of the paper discussed above.
In one case, two well regarded researchers are suggesting that such tracks can *only be accomplished* after young inexperienced birds have used their innate migratory heading to travel the standard migratory route first time round. Then the global co-ordinates of that migratory route that they have recorded, can be downloaded into their innate GPS guidance system packages in order to activate them.
The following review paper, also published by the Company of Biologists, sets out this proposition and appears to be a second example of switching horses in mid-stream.
Note. K.Thorup has previously published the results of many unconfirmed Emlen-caged orientation experiments.
"The bird GPS – long-range navigation of migrants, K.Thorup and R.A. Holland, JEB.,212:3597-3604 2009.
http://jeb.biologists.org/content/212/22/3597.full

The paper reviews evidence going back as far as 1952 and the summary, with key points underlined, reads thus;.-

Nowadays few people consider finding their way in unfamiliar areas a problem as a GPS (Global Positioning System) combined with some simple map software can easily tell you how to get from A to B. Although this opportunity has only become available during the last decade, recent experiments show that long-distance migrating animals had already solved this problem. Even after displacement over thousands of kilometres to previously unknown areas, experienced, but not first time migrant birds quickly adjust their course toward their destination, proving the existence of an experience-based GPS in these birds. Determining latitude is a relatively simple task, even for humans, whereas longitude poses much larger problems. Birds and other animals however have found a way to achieve this, although we do not yet know how. Possible ways of determining longitude includes using celestial cues in combination with an internal clock, geomagnetic cues such as magnetic intensity or perhaps even olfactory cues. *Presently there is not enough evidence to rule out any of these, and years of studying birds in a laboratory setting have yielded partly contradictory results. We suggest that a concerted effort, where the study of animals in a natural setting goes hand-in-hand with lab-based study, may be necessary to fully understand the mechanism underlying the long-distance navigation system of birds. As such, researchers must remain receptive to alternative interpretations and bear in mind that animal navigation may not necessarily be similar to the human system, and that we know from many years of investigation of long-distance navigation in birds that at least some birds do have GPS – but we are uncertain how it works.*

Comments on the four underlined sections.

First underlined section.

Only two cases of displaced adult long-distance migratory species successfully homing to their winter quarters (and first-time migrants failing to do so) are provided in the table of examples. One refers to the K.Thorup et al. (2007) displacement experiment where adults were released in the grounds of a University, the juveniles at an airport (see Appendix 3), and no controls were released to establish the "normal" winter location. Even if they had "quickly" adjusted their course (to this unconfirmed location), this is no proof of possession of a GPS. In a later paper co-author R.A.Holland (2014) defines "quickly" as being *within the first 100 kms of departure from the site of displacement.*

The other case mentioned, refers to Perdeck's 1958 experiment detailed in Appendix 4. Perdeck's adult starlings most certainly did not quickly adjust their course to their destination (Appendix 4, Figure 2 and Gerrard 1981a,b), yet Holland states *"The clearest example (Perdeck 1958) demonstrated that adult but not juvenile birds are capable of migratory true navigation."*

Second underlined section.

Determining longitude with an internal clock … or with the aid of Earth's magnetic field…. or by smell? - see Chapters 3 & 4.

Third underlined section.

researchers must remain receptive to alternative interpretations. Very sensibly put, but they then go and spoil it……..

Fourth underlined section.

we know from many years of investigation of long-distance navigation in birds that at least some birds do have GPS

So there can be no ambiguity as to what readers might think the authors mean by "GPS", they set this out in a section of the publication describing navigational terms:-
"Global Positioning System. A satellite system enabling the determination of ones location (latitude, longitude and altitude) with an accuracy in the order of less than 10 m anywhere on earth using a GPS receiver".

"some long-distance migrant birds do have GPS – but we are uncertain how it works" ! As GPS relies on man-made Earth orbiting satellites and birds were "navigating" long distances quite successfully before *Homo sapiens* placed satellites into orbit, either birds launched their own miniscule satellites thousands if not millions of years ago, or the conventional world of experimental biology is in chaos.

Appendix One

More directional influence questions and explanations

A continuation from Chapters 6 and 7.

To what extent do climatic factors influence migrant routes?
In places where the prevailing wind comes in off the ocean, as in the case of the western British Isles, it often brings with it moist warm air. It is usually only when a high-pressure system becomes established in such areas, bringing with it colder clearer air from the north, that many potential autumnal migrants are finally triggered into a move. So without a full understanding of local conditions, one might be led to think that autumnal migrants generally leave the British Isles in adverse conditions.

Ocean currents have an affect on the location of migrant seabird and wader species whenever the birds are resting on the surface. This will have little influence on destination, except in the case of penguins. For the multitude of species which winter out on the open ocean, winds and currents will be a dominant factor in the changing of location.

What effect has topological/topographical memory or recognition upon migrant routes?
In some species it is fairly easy to see how topological recognition can help, year after year. In others, that wander over considerable distances in the winter, and/or breed in differing localities from one year to the next, it is far less likely to be of much assistance.

As an en-route influence topological/topographical recognition and memory can have far-reaching effects. Topological memory obviously only influences individuals that have either travelled the route before, or are in the company of those that have.

Exactly where is the line drawn between the alternative diurnal and nocturnal migratory behaviour?

There has traditionally been some confusion over how fixed in their ways migrants really are, and the information now forthcoming from tracking devices demonstrates the adaptability of many migratory travellers. Too much to expect birds to switch readily from one highly complex navigational method to another? With the pecten aided innate directional response in the frame, the switch is automatic, requiring no advanced navigation techniques.

Starlings, typically early morning diurnal migrants, will shift location during the night, possibly encouraged to do so by a sudden drop in temperature, or disturbance. Radar evidence was mainly responsible for bringing this to light with the aid of the starling's recognisable flight signature.

Observer evidence has proved that chaffinches sometimes commence autumn movements long before dawn. In coastal areas many migrants follow a coastline during daylight. Especially in spring, nocturnal migrants are known to travel long distances, usually early in the morning or evening, possibly sent on their way by aggressive local birds, lack of food or shelter, or simply because of glandularly induced continuing urgency.

Of course there are times when the local conditions force an alternative, as for example, when a nocturnal passerine migrant finds itself out over the sea at dawn and still out of sight of land; in extreme cases where there is no darkness (Greenland in late May); when a diurnal migrant reaches an insurmountable barrier such as a coastline that can perhaps only be overcome when it is no longer such an obvious deterrent (after dark). All these variables will affect the direction of migration routes.

As many nocturnal migrants are known to carry large fuel reserves that will enable them to fly non-stop from dusk to dawn (if not longer) why do they often call a halt in the middle of the night?
Radar observations have shown that most nocturnal migrations are terminated long before dawn and that the total volume of migrants diminishes rapidly after midnight; the participants either landing or reducing altitude with the intention of doing so. The absence of further directional cues (cloud cover preventing continuing star following for example) could also be a contributing factor. Individuals who call an

early halt in crosswind conditions (Chapter 6, Figure 6.9) may be at an advantage.

However, many small migrants put on large amounts of body fat prior to undertaking serious migration and will then cover amazing distances in one burst before arriving at the next suitable refuelling location. A relaxing stop of several days or even weeks, and then another sudden long flight. Such individuals tend to undertake these stop-go leaps in good weather conditions – tailwinds and clear skies. This regularly recorded stop-go style of migration would play havoc with almanac updates!

Why do birds migrate at different altitudes?
Birds tend to remain close to Earth's surface when flying into wind, and yet often fly very much higher in tailwind conditions. This is of course the most efficient way for a bird to get about, the wind being weaker close to the ground. Without becoming too involved in flight mechanics, a study of fledgling behaviour shows that landing into wind is an acquired art, often painfully learned at an early age when downwind crash landings and low speed stalls are frequent.

Early understanding of the relationship between air speed and ground speed is essential for survival and is partly learned and partly innate, as anyone who has raised an orphan wild duckling will confirm. Extremely confused whilst running and flapping after an imprinted wellington boot when a gust of wind suddenly takes it aloft. Yet in its element when placed in a tidal pool, diving and swimming far beneath the surface, never having been anywhere near water before.

If a bird is battling a headwind whilst heading for a ground-based objective, yet no real headway is being made, one of two alternatives will permit the eventual arrival. Either the flying speed must somehow be increased, or a temporary halt must be made (by first decreasing altitude). In doing so the wind speed will probably decrease and the destination be reached at a lower altitude.

Reversing the situation; if a migrant is flying downwind, the ground speed will appear faster than normal and the bird can comfortably afford to gain altitude (for a better view of the terrain or whatever). This in turn will speed progress further because with altitude increase (usually) comes higher helpful wind speed. And if the higher altitude winds are not helpful, the terrain beneath will appear to slow or be slipping sideways and the bird may well be induced to reduce altitude.

If bird navigation really is such an haphazard instinctive affair, why are there not more lost and wandering migrants about?
Millions of migrants (nearly all small passerines) die each autumn. In many instances this is simply through being poorly prepared for the rigors ahead – natural selection at its most cruel. Once off course for whatever reason, (but not because one of their compasses was faulty surely?) and thus outwith an area that permits continuing survival, they drown or starve and are rapidly consumed by predators. Either way they are not readily noticed.

How do juvenile migrants know when to commence autumn migration?
An autumnal migratory move to some other location permits survival and although adults might conceivably be aware of this, juveniles cannot be. They may simply join adult flocks (as with most geese and swan species) but are often abandoned and left to fend for themselves. In some cases adults depart before their offspring are even fully fledged. Clearly whatever the underlying urge to move, it must be considerable in order to force a youngster to leave its comfortable habitat in favour of the wide blue yonder.

From the day of the summer solstice, each day becomes shorter until the night of the winter solstice 6 months later. The immediate effect of this is that in many species there is less time for feeding, less exposure to sunlight, and longer roosting periods. With lengthening nights comes a general drop in temperature that also affects the food chain. The avian biological clock that is regulated by surrounding seasonal changes, triggers seasonal moult, breeding desires, nest-building and everything else that makes a bird tick. Circadian rhythms; the *annual* clock that helps force an unwilling bird to depart induces agitation; commonly referred to as "migratory restlessness". Food is more readily converted into fat, that will later be used as "in-flight" fuel. In non-migratory species that possess a wider ratio of acceptance, these two reactions would remain un-triggered or would require a much greater changing habitat jolt.

Generations of selective elimination of avian non-conformists has usually insured that the innate internal clock triggers the move before the nights get too cold and before the food runs out. A sudden drop in temperature is often caused by clear night skies and in turn this usually occurs in areas of higher pressure and favourable wind systems. Conversely low pressure often brings humidity, overcast conditions and adverse winds.

How do migrants know when to make en-route stops?
Having departed, hopefully with others, the group will usually spend a considerable time in cheeping, tweeting, honking or hooting at each other by way of encouragement and flock cohesion. They will stop again when they become tired and when a suitable resting area is reached. If adverse weather conditions are encountered or need to refuel forced the stop, the next leg of the journey may be delayed.

In cases where a long crossing over apparently inhospitable terrain is faced, that innate trigger kicks in again. Natural selection demands a long refuelling stop and special attention to the strength and direction of the wind. After a successful and often very long crossing, another long refuelling stop (sometimes lasting weeks) is usually required.

How is the spring return migration triggered?
Because the autumn "shut-down" of higher latitudes generally commences first, such migrants are actually overflying others still in two minds as to when to pack their bags. Those already oriented in the right direction must surely influence these later starters.

But in the spring, that influence is usually absent because it is often the shorter distance migrants that will be heading for their lower latitude breeding grounds first. One of the guiding factors on offer now would seem to be knowledge or recognition of ground covered the previous autumn "on the way down". At least the novice migrant is now far more knowledgeable than heretofore. However as most will have wandered around their wintering area, something more by way of a signpost is required.

Tracking of adults in similar circumstances suggests that even experienced individuals have difficulty deciding on a migratory direction and are likely to initially depart further south by mistake. But this will soon put them in direct competition for food with sedentary tropical species in their breeding season. It may also bring them into contact with southern hemisphere breeding species moving up into "their" wintering areas. Finally, a few may be influenced by overflying species that have wintered far into the southern hemisphere.

How is the spring return finally halted?
In some species the spring arrival in a suitable nesting locality is achieved well in advance of any actual breeding activity, and this is especially noticeable in species that take a number of years to mature. The inference is that such birds arrive earlier for a very good reason. In the case of the sooty shearwater for example (Chapter 8); if these arrived later they would either arrive elsewhere or discover all the best nesting sites were already occupied.

Why are some species migratory, yet other closely allied species mainly sedentary, and why are some individuals migratory and others not?
A great deal depends on the prevailing climatic conditions existing over the breeding range of any species prone to migratory hesitancy. If food was abnormally plentiful and the weather unseasonably mild, those factors might override or delay the innate urge to depart.

What is the difference between cold weather/warm weather movements and true migration?
The direction of birds involved in weather movements (including temporary reverse migration at times when unacceptable conditions are suddenly encountered) is naturally variable (Chapter 8). However once a shift is forced in a manner that results in a major displacement, the movement could become a forced "escape", in which case phototactic influences could dominate.

What are the differences between juvenile dispersal movements and true migration?

True migration is basically an innate overriding urge to shift. On the other hand juvenile dispersal would seem to be nothing more than a bemused wandering in or through survivable habitat. Such wanderings would tend to be lateral within that suitable habitat band but would leave the participants well to the west or east of the species' normal migratory departure region (Chapter 10, Figures 4 and 5).

How do first-time migrants know in advance they are about to embark on a long flight across inhospitable terrain?
Logically the answer must surely be – they cannot know in advance and the numbers of juveniles that fail to make a crossing of a major desert or ocean must be enormous. The minority that survive will hopefully remember the experience in future.

However there are a few pointers. On reaching the edge of an obviously hostile environment most migrants of all ages, will stop and refuel. If no obvious café, they often back-track until they find one. Being in the company of possibly thousands of others of numerous species, young and old, all behaving likewise, offers a better chance of setting out carrying a full fuel load. In any case hundreds or thousands of years of rapid and drastic selective weeding out of the weak has already left a strong gene pool – and as (for example) the Sahara expands in size, that pool may be stretched to its limits – and beyond. In which case the few survivors of that particular species will either have stopped their migration and wintered in northern Africa or successfully overflown the widening gap. One group will quite rapidly become a smaller and less successful sub-species, and the other a longer winged or larger sub-species capable of carrying more fuel. Migratory evolution in progress once again.

Most species of birds do not in fact possess eyes situated squarely on either side of the head, so how does this affect the "right-angles to low-angle solar glare" orientation?
Only when the pecten's position in the avian eye of each species and it's influence can be properly diagnosed, will a full understanding as to how the glare influenced migratory headings will differ from one species to the next. A long job for experts if any can be persuaded to investigate.

For example, although suggestions have already been advanced to explain why birds should be forced into one-eyed responses to low level solar glare at times of phototactically induced movements, the actual orientation adopted at such times (eye angles or heading angles) is based mainly on the known headings of diurnal migrants at such times. But in some cases up to 45 degree divergence would fit the range of the known performances more accurately and this is possibly because the eyes are placed on the head so as to afford all round vision as in the case of the Woodcock.

http://www.woodcockwatch.com

A number of species migrate with the aid of thermals, and whilst gliding round in circles within these rising warm air pockets, would not be influenced phototactically. Likewise owls and other birds of prey with wide binocular vision. Both groups would either be noted for a lack of migratory enthusiasm, be simply following in the wake of their prey (which would be migrating "normally"), or be within a group of experienced adults.

Many other situations have never been properly investigated and until the pecten puzzle has been addressed, serious questions will remain partially unanswered. Two cases are discussed in Appendix 2 and Appendix 3.

Appendix Two

Pigeon homing in World War Two

The pecten structure hypothesis implies flight headings across featureless terrain should be longitudinal, not latitudinal.

Release a trained homing pigeon over the ocean well to the west or east of its home loft and far out of sight of land and it should be unable to home directly.

World War Two provided the most extensive experimental platform for the study of long-range pigeon homing achievements one could wish for. Whereas normal competitions usually involve hundreds of simultaneous releases in reasonable weather conditions (and individual performances are influenced by others), the wartime releases were either individual affairs or in pairs.

As many UK aircraft with space available carried one or two birds on flights, there were naturally plenty of releases from all points of the compass. Additionally many birds were released in France, Belgium and Holland by people who were engaged in clandestine operations, and by the allied armies. Nearly 200,000 young birds were donated to the Armed Services in Britain during the war. Whilst accurate record of the number of releases was obviously not possible, most *successful* homing achievements were noted. So whilst one cannot say specifically that a certain percentage of releases managed to home within a certain time, one can draw some negative conclusions.

There seems to be no evidence of any homing achievements from *east* or *west* when released clearly out of sight of land, despite the fact that there must have been hundreds of releases in such conditions. Random dispersal and/or downwind drift, should by the law of averages, have produced between 10 and 50 successful homing feats from east or west in this period. This not being the case suggests that the birds in question were indeed forced into innate behavior that resulted in suicidal flights in wrong directions nearly every time because of the north/south pecten structure influence (Chapter 6 Figure 6.7).

The following quotations, uplifted from the only collective record of homing feats relating to WW2, *"Pigeons in World War Two, W.H.Osman, 1950"*, that singled out exceptional feats (some of which *appear* to have latitudinal homing components), is illuminating.

"NPS.42. 22876. On 16th June 1943, this pigeon was released from a dinghy in the Atlantic, approximately 100 miles from the coast, and homed to base in 1 hour 10 minutes with its message. Trained by RAF Station Pembroke Dock."

A good example of west-east homing from out of sight of land? However - and no discredit to the pigeon, its handlers or the authors – once one examines the case in detail the bird either homed at approximately 154 mph or it was in sight of land to the east from lift off.

"NPS 42.30687. On 23rd March, 1943, was released from an aircraft 80 miles out in the North Sea at 14.55 hours. Delivered its message at 16.50 hours (80 miles in 1 hr. 35 mins.). Bred and trained by RAF Station North Luffenham."

"NPS 42.30704. On 23rd March, 1943, was released from an aircraft 80 miles out in the North Sea at 15.00 hours. Delivered its message at 16.28 hours (80 miles in 1 hr. 28 mins.). Bred and trained by RAF Station North Luffenham."

Possibly two east-to-west homings (? from the same aircraft) from just out of sight of land, but Luffenham is 35 miles inland from the nearest point of the North Sea. The wording is a bit misleading but it is significant that these relatively easy returns were singled out for mention.

"NPS. 43. 4112. During the evening of 16th September 1944, was released from a dinghy of a ditched aircraft in position 49.15N.,04.33W., about 100 miles from base, with an S.O.S. message which it delivered next morning. The crew were rescued. Pigeon arrived wet and smeared in oil. Trained by RAF Station, Mount Batten."

The given position is exactly 79 miles from the RAF station and the nearest point of land is about 57 miles to the north of the ditched position and within sight. The bird would surely have dried out during its flight?

"NURP.41.GMN.199. On one of her 64 operational sorties, this pigeon was air-released from a Beaufighter over the Bay of Biscay 200 miles from base. The Beaufighter is a difficult aircraft from which to safely release a pigeon and this bird was evidently injured in the release. It managed to reach South Wales and was picked up in Llanelly on the same day, whence its message was transmitted by the police. Trained by RAF Station Chivenor."

200 miles from RAF Chivenor is not in the Bay of Biscay and nearest land (probably the Scilly Isles) would have been about 70 miles distant and within sight at altitude.

Note. Appendix 2 is a précis of Appendix 1 from "Instinctive Navigation of Birds", 1981 by the author.

Appendix Three

Mallard nonsense orientation

The experiments that created the expression *Nonsense orientation*, and to this day stimulates debate were conducted by G.V.T. Matthews *(Bird Navigation, Cambridge University Press, 1968)*.

Matthews transported large numbers of mallards *(Anas platyrhynchos)* from Slimbridge Wildfowl Trust in Gloucestershire to various parts of southern England, where they were released, one at a time, in sunny conditions, at most times of the day and more or less at all times of the year. Their initial departure direction tended to be between north and northwest time and again...why?

Certainly this was not the direction that most of them would have chosen had they wanted to get back to Slimbridge (A.3 Figure 1). Neither was it in the direction taken by migrating mallards.

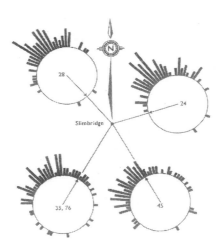

Figure A 3. 1 Nonsense orientation of released mallards.
The shortest spoke represents the vanishing bearing of one bird.
Direction and distances in miles of release points from Slimbridge are shown within each circle.
Source: Derived from G.V.T.Matthews (1968).

In order to forestall any criticism of topological or other similar influences which might have produced such results (prompted by having some previous experiments questioned in this way), Matthews stated *"such proponents of topological orientation would have to explain what features were common to all 15 points at which northwesterly orientation of mallards was observed to give rise to such an orientation."*

He then mentioned *three* topological features that were common to all release points, and there was a fourth buried in the mass of technical data provided. The three common features were as follows; with bracketed comments:-

"Most of the release points were chosen so that any mallard country lay to the south and east..." (Why?).

"Additionally it was arranged that any rising ground should be to the north and west of the release point" (Why?).

"Nearby water must be avoided as the birds alight on it. Deserted airfields made ideal release points." (This was fair comment, except for the fact that, as all British pilots know, main runways in the United Kingdom all possess the same directional bias because of the prevailing wind).

In fact all the deserted airfields used by Matthews had the main runway lying more or less on an ENE/WSW axis, and indeed Shobden, where for some reason 9 of the 27 individual experiments were conducted, had only one single huge runway lying on an east-west axis. There was always higher ground to the northwest (Figure A3.2), although in a few cases this was marginal by human standards, and at one site where two experiments were conducted there was considerably higher ground in another direction.

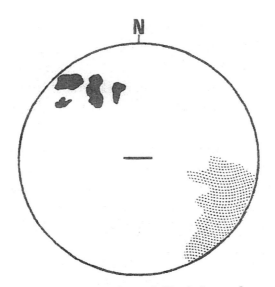

Figure A 3. 2   Matthews' Shobden release site.
High ground over 350 meters to NW.
Low ground below 100 meters to SE.
Note E-W direction of runway; center.
Radius of circle; 12 kilometers.
Source: Gerrard (1981a).

This evidence is enough to make the whole vast experiment rather pointless at first glance, especially when one considers the fact that Kramer (1959) had already published his comments on initial orientation in relation to "familiar looking' home surroundings

One could probably account for the northwesterly orientation by suggesting that the released birds initially headed for the nearest higher ground …certainly they were more than familiar with such features at Slimbridge. However, at least 838 out a total of 868 releases were made in sunny conditions when the Sun was in the *southern half of the sky*. Out of the 868 releases, only 517 were actually recorded as having departed in the direction between north and west (apparently from 14 different release sites, not 15). 195 birds moved off in other directions, and the remaining 156 did not, it seems, move off in any direction sufficiently specific to record.

So some 60% of releases simply flew towards the clearly visible higher ground in the general direction that afforded a good clear view of said higher ground and *away from the sun arc*. In almost as many cases, one could offer an equally good alternative explanation by stating that the birds departed by flying directly away from the runway in the general direction that afforded good clear visibility (regardless of where the higher ground was situated).

After all, one should not be surprised to find releases from the side of a motorway (avian or human) flying off into the fields. However this latter behavior is still liable to be influenced by "attractive" terrain whenever this is visible, so if "the-away-from-the-runway" response was the sole directional trigger one would have expected far more S/SE departures (i.e., flying off the other side of the runway: especially when the Sun was fairly high in the sky.

Several other experiments by Matthews tend to reinforce the contention that the initial departure orientation was caused by attraction to "familiar" type objects, aided by the angled sunshine making these objects obvious. He obtained an indication that mallards exhibit these directional departures at night whenever the level of visibility is reasonable (clear skies, with or without a moon) and that different stocks of mallards exhibit different directional departure responses, doubtless caused by different imprint attraction (such as attraction to low ground when used solely to such, to houses or buildings when especially familiar with built-up areas and so on).

Therefore these free-flying mallard releases exhibited exactly the same initial orientation responses as Kramer's untrained homing pigeons just mentioned. Matthews' nonsense orientation ability claim appears completely explicable and very far from nonsensical.

Note. Appendix 3 was first published in Chapter 12 of "Instinctive Navigation of Birds", 1981 by the author.

Appendix Four

The Perdeck Saga

Introduction.

A.C.Perdeck trapped and ringed (banded) over 19,000 starlings in Holland during the months of October and November 1949-1957, and released almost 11,500 of them in Switzerland. The remaining 7,500 were released in Holland to serve as controls in order to verify the normal onward direction and the true wintering area of the birds being trapped and displaced.

The results of this massive and costly experiment were published in 1958 and caused a sensation because Perdeck came to the revolutionary conclusion that during autumn migration, adult starlings used a true goal orientation (homing orientation), the juveniles a one-direction orientation. Starlings were able to fix their winter quarters in their first year, with an ability to reach it in later years by means of true goal orientation (pages 33-34 of the published results).

So, whilst the underlying direction of the migration route was *innate* within a species, the ability to get back on course when displaced laterally during migration was an acquired (*learned*) art. Hence the comments on this claim in the first chapter of this book…..”Wow”.

Figure 1.1 and Figure A.4 1 outline this claim.

The full details of A.C.Perdeck's massive displacement experiment can be viewed at the publisher's web site - 38 pages of somewhat confusing data presentation, as we shall see.

http://www.nou.nu/ardea/ardea_show_article.php?nr=1562

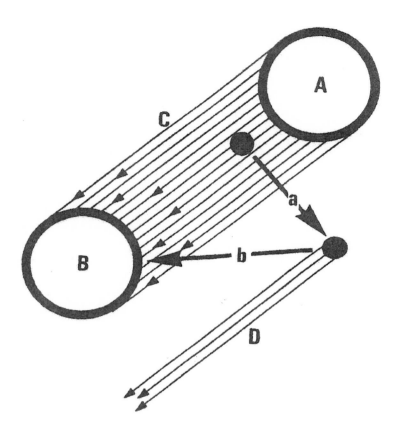

**Figure A 4. 1   Outline of Perdeck's starling displacement
hypothesis.**
A. Breeding area.   B. Normal wintering area.
C. Preferred direction.   D. One-direction orientation of juveniles.
a. Lateral displacement from Holland to Switzerland.
b. True-goal orientation of adults to winter quarters.
Source: Gerrard (1981a)

Virtually every article involving the subject of avian migratory
navigation mentions details of Perdeck's displacement experiment,
however briefly, and all too often in support of some other experiment
or navigational claim. In part, because of the confusing presentation of

data, many of these articles manage to misquote.

A selection of straightforward misquotations.
Note. Every one of the following quotations is incorrect, even if we accept Perdeck's published results as proven facts.

J.Dorst, The Migrations of Birds, pp. 331,1962. *"Adults... returned to their customary winter area"*.

Edwards (Producer), Bird Brain,: The mystery of Bird Navigation, BBC TV documentary, 1974. " *.....most adults were found in their normal wintering area.*"

In both these cases only 10 adults were found in their winter quarters, 56 were located elsewhere, and the whereabouts of the other 3,812 was unknown.

Fitter (Ed.), Book of British Birds, pp. 318, 1969. *"Many of the older birds, which had visited the winter quarters at least twice before, corrected for their displacement inland by flying northwest.*
At least twice before?

R.Baker (Ed.), The Mystery of Migration, p 163, 1980. *"The adults ....took a northerly direction to the wintering area."*
Only 5 out of 3878 did so.

As mentioned in the Preface, late in 1970 NASA joined forces with the Smithsonian Institution and the American Institute of Biological Sciences, and sponsored a symposium on Animal Orientation and Navigation at Wallops Station, Virginia. The list of names attending was impressive.

In session 4: Bird Migration and Homing, 2 papers were delivered on subjects that could feasibly refer to Perdeck. Both did, and both managed to misquote.

S.T.Emlen of Cornell University delivered a paper entitled "The Ontogenetic Development of Orientation Capabilities", the research for which had been assisted by grants from the US National Institute of

Health and the US National Science Foundation.

In short, Emlen was attempting to find out the equivalent of how Perdeck's displaced adult starlings could navigate back to their winter quarters whereas juveniles could not, via experiments with a different species constrained in artificial planetarium conditions. His apparatus has since been shown to possess serious faults (Chapter 1).

Early in his delivery he stated – *"Actually field experiments conducted over a decade ago suggested a dichotomy of navigation capabilities between young and adult birds. When birds of <u>several species</u> were captured and displaced from their normal autumnal migration routes, the adults appeared to correct for this displacement and returned to the normal winter quarters while immatures (birds on their first migration trip) did not, but rather took up courses parallel to their original direction of migration. This implies an improvement in navigation performance as a result of previous migratory experiences. I have arrived at a somewhat similar conclusion from studies of the migratory orientation of caged indigo buntings."*

In the printed text of Emlen's paper he mentions *"references 5-8"* in bibliographic support of this statement, which in turn refer to experiments by W.Ruppell and F.Bellrose as well as to Perdeck. But the experiments of Ruppell and Belrose did not find Perdeck's claimed difference in navigational behavior between adults and juveniles.

At the same session, E.Gwinner of the Max Planck Institute delivered a paper entitled *"Endogenous Timing Factors in Bird Migration"*, the main theme of which suggested support for the hypothesis that endogenous (internally formed) factors are involved in the termination of fall (autumn) migration in first year warblers. In endeavoring to widen the scope of his experimental work with warblers so as to include other species, he drew on the field displacement experiments of Ruppell and *Perdeck* for support. Because Gwinner was only interested in the behavior of displaced *juvenile* migrants, the apparent differing behavior of Perdeck's adults did not concern him and consequently he did not produce the usual type of quote-misquote. Instead what he said of Perdeck's displacement experiment was this- *"These (juvenile) birds that had started migration comparatively recently continued to travel in their original direction, even if the environment of the release point was favourable for wintering. Only those birds which at the time of capture had already almost terminated fall migration stayed in the vicinity of the release site, provided the release site was in a favourable environment."*

Pure speculation as to which stage of migration the juvenile displaced birds had reached.

By the late 1970's the main focus on the avian migration "puzzle" had switched from the use of solar or stellar compasses (the first "2 C's") to magnetic ones (the "3rd C"). The thinking behind this switch appears to have been prompted by the realization that many western European migrants seemed to be very roughly following fixed magnetic headings. Perdeck's *displaced juvenile starlings in particular.* A single publication has been selected to emphasize this point.

In 1991, Berkhauser published a book *"Orientation in Birds"* (reprinted in paperback form in 2013 and available from Amazon) with a forward by Rudiger Wehner and edited by P Berthold.

The behaviour of Perdeck's displaced *juvenile* starlings featured prominently in four chapters written by five leading German researchers. Each of the chapters gave a slightly different version of Perdeck's claimed results.

The 4 chapters.
1 Hans G. Walraff *(Max-Planck-Institut fur Verhaltens-physiologie, D-W-8130 Seewiesen Post Starnberg, Germany)* contributing the chapter *"Conceptual Approaches to Avian Navigation Systems",* makes the following opening statement on page 128;
"The general basis of migratory orientation in birds is most probably an endogenous time-and-direction programme....."

On page 129, referring specifically to Perdeck's 1958 paper, Walraff states:-
" Young Starlings, migrating for the first time, when displaced perpendicularly to the compass direction normally taken by the population, continued to fly this normal compass course and hence arrived in an abnormal area dislocated by approximately the direction

and distance of displacement".

On page 130, a diagram (Figure 1, reproduced here A 4 Figure 6 below) confirms the above statement showing 2 similar areas, suitably dislocated, and the text beneath reads:

"Starlings migrating in autumn through Holland (H) arrive from breeding grounds in northeastern Europe and subsequently winter in southern Britain, northern France, Belgium and Holland, in an area approximately limited by the solid line surrounding the hatched arrows. Many such Starlings were displaced from Holland to Switzerland (S) and released there. In subsequent winter months, most of the juveniles (empty arrows) were found in an area displaced accordingly (as marked by the broken line). ...." Schematized by Walraff (1974) after Perdeck 1958.

This is followed by:-

"Bearing-and-distance migration. Compass orientation. The result obtained with the young Starlings suggests the conclusion that some environmental references were available in Holland as well as in Switzerland, according to which an identical compass course could have been chosen." ..... "Hence the geomagnetic field fits the above requirement in an almost ideal way."

And on page 134:-

" Intended direction. To explain the behaviour of the young Starlings in Perdeck's experiment (Figure 1) it is not sufficient to determine the directional references which must be available in Switzerland as well as Holland. The birds apparently followed some 'internal command' to select just one specific angle - an intended direction - with regard to these references. What is the source of information telling the birds which angle they have to adopt and maintain? Empirical research strongly suggests that intended migratory directions of juvenile birds are products of evolutionary processes, during which most appropriate directions for the various populations have been selected and genetically fixed (Helbig & Wiltschko, 1987; Berthold 1991; Helbig 1991.

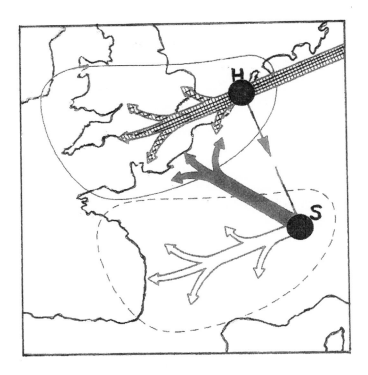

Figure A 4. 2 Walraff's Figure 1 drawing based on the recapture areas of Perdeck's juvenile displaced starlings.
The solid oval outline represents the normal wintering area.
Source: Derived from H.G.Walraff/P.Berthold (1991).

Author's note.
The average distance of migration of the displaced juveniles is in fact *double* (as is obvious from Perdeck's diagrams) and the average direction is *different by some 20 degrees* as Perdeck goes to some trouble to point out on page 16 of his own publication (see case 3 below for the actual wording).

2 A.J. Helbig *(Institut fur Pharmazeutische Biologie, Universitat Heidelberg, Im Neunheimer Feld, 364, D-6900 Heidelberg, Germany)* contributed a chapter *"Experimental and Analytical Techniques used in Bird Orientation Research".*

On page 279 he wrote:-
"The displacement of marked birds prior to or during migration provided strong evidence for the long-held assumption that directions of migration in young birds are innate, i.e. based on genetic information."

And on page 280:-
"Perdeck's (1958) results with Starlings were especially impressive: he transported over 10,000 birds (adult and juv.) during autumn from Holland to Switzerland, i.e. perpendicular to their normal migration route. The large number of recoveries of displaced birds showed that the adults attempted to compensate for the displacement, and flew towards their normal winter quarters in southern Britain and northern France. The juveniles, on the other hand, continued in the same direction they would have followed if not displaced: most were found southwest of the release site in southern France and Spain."

Author's note.
The misquotation regarding the subsequent directional behavior of adult displaced birds has already been mentioned. The majority of displaced juveniles *did not* continue in the same direction they would have followed if not displaced. Most were *not* found southwest of the release sites in southern France or Spain; only 71 of the 171 recaptures were found in this area.

3 W. and R.Wiltschko *(Universitat, Zoologie, Siesmayerstrasse 70, D-6000 Frankfurt a.M.,Germany)* in a chapter *"Magnetic Orientation and Celestial Cues in Migratory Orientation"*, referring to Perdeck's 1958 displacement experiment paper, state on page 17:-
"Thousands of transmigrants were caught, and transported at right-angles to their normal migration route to Switzerland, where they were released. Ring recoveries disclosed their later whereabouts. The majority of the young birds' recoveries came from the southern French Atlantic coast and northern Spain, indicating that these birds had continued on the west-southwesterly course that had brought them to Holland - and which under normal circumstances, would have brought them to their traditional wintering range".
"These findings .... suggest that young birds on their first migration fly fixed courses. The innate information they possess on the position of

their species' wintering range seems to be given in "polar coordinates", namely as a direction and a distance to be travelled, with the distance controlled by an endogenous time programme and the amount of migratory activity.

However Perdeck had written:-
"The main direction (referring to the onward direction of juveniles displaced to Switzerland*)* *"....appears to be SW by W (236°)"* and...*"This direction is close to the broadfront migration over the Netherlands, which was concluded to be between W & WSW (between 247° & 270°) (see p 8). There is however a difference of some 20 degrees."*

Author's note.
The majority of young bird recoveries *did not* come from the southern French Atlantic coast and northern Spain; only 76 out of 171 did. They had *not* continued on the west-southwesterly course either, as Perdeck himself had pointed out.

4 P.Berthold, the book editor *(Max Planck Institut fur Verhaltensphysiologie, Vogelwarte, Schloss Moeggingen, Radolfzell, Germany)* contributes a chapter titled *"Spatiotemporal Programmes and Genetics of Orientation".* On page 91 he writes;
"Young European Starlings (Sturnus vulgaris) trapped in the Netherlands during their autumn migration from the Baltic region to west European wintering areas and transferred to (and released in) Switzerland, continued their autumn migration to Spain, an area that normally is not reached. Thus, they continued their journey in the programmed direction, and to some extent also for the expected distance, in spite of the transfer (Perdeck, 1958).

Author's note
The displaced juveniles did *not* continue their journey in the programmed direction and they travelled on average, *twice as far.* Perdeck clearly showed in his Figures 8 & 11 that the onward distance travelled of the displaced juveniles was double and indicated the exact area of each recovery. Only 40 of the 171 recaptures had continued their journey to Spain.

The *4 different* misquotations of Perdeck' published displaced juvenile onward directions in the same book.

	Through Holland	On from Switzerland
Perdeck.	**Between W & WSW**	**SW by W/20° less**
Walraff	Betwn W & WSW	Betwn W & WSW (wrong)
Helbig	SW (wrong)	SW (wrong)
Wiltschkos'	WSW (wrong)	WSW (nearly right)
Berthold	Betwn W & WSW	Betwn, W & WSW (wrong)

The five contributing experts in this *"Orientation in Birds"* book have published experimental claims of their own that separately underline their collective contention *that young birds inherit the ability to migrate in a specific direction and for a set distance.* These dual beliefs were developed more or less independently, hence the four slightly *different* interpretations of the same *"inherited"* genetic theme in the four different chapters. The underlying mechanism for maintaining direction was four to one in favour of some sort magnetic compass.

The reasoning behind the experiments.

In the introduction to Perdeck's published results, we are told that there was, at the time, no firm evidence in support of en-route lateral "get back on course" compensation by avian migrants, *of any age.*

But unfortunately the experiment was being designed to attempt to confirm a two part hypothesis, because the organizers go on to write that although due notice should be taken of adult birds, *a difference in the behavior of adult and juvenile birds could be expected.*

Why should a difference be *expected*? If it could be discovered that adults can home in on their winter quarters after lateral displacement by some miraculous means, why not juveniles also?

On page 4, whilst still discussing the planning of the experiment, is the following logical proviso:-
  "If (the adults) are using true-goal orientation, they will go straight to their resting (wintering) area in a different direction to the original course." Note the words *"they will go straight to"*.

It is obvious what prompted the Vogeltrekstation Texel Foundation to sponsor this *joint* investigation. First, there were tens of thousands of easy to catch noisy robust starlings migrating annually through Holland and second, the joint hypothesis, however outlandish, should be easy to prove or disprove. Catch loads of adult and juvenile starlings (easily separated by plumage differences), place an identifying ring (band) on a leg, transport them at right-angles to their normal migration route to an area they would be unfamiliar with, and sit back for the ringing results to come flooding in.

The published results duly claimed to have demonstrated that both parts of the hypothesis had been verified. *Adult* birds could somehow counter the effects of lateral displacement and fly *"straight to their resting (wintering) area in a different direction to the original course."* (the stated conditions for verification of part one of the hypothesis), whereas *a difference in the behavior of adult and juvenile birds* was indeed demonstrated (the stated conditions for verification of part two of the hypothesis).

In fact displaced adults had not demonstrated an ability to fly *"straight to their resting (wintering) area"*. The results did however confirm that the juveniles tended to behave differently. But why they behaved differently was not properly investigated because attempting to make two new discoveries in one massive experiment had thrown a spanner in the works from the outset.

The flaws in the claims.

In order to establish that *adult* migratory starlings can "home" onto their normal winter quarters the normal wintering area (the boxed sector in Figure A. 4) of the *adults* has to be confirmed, which can only be done by releasing *adult* controls. None were released, all 7,500 controls being juveniles. This was because about four times more juvenile than adult starlings were being caught and in order to test the prime hypothesis, it was decided to transport all the adults to Switzerland (about 3,900 of them) along with the other half of the juveniles (about 7,500).

From previous ringing returns the organizers knew where many of the adults starlings migrating through Holland would be wintering anyway; they were probably right. It made little difference because none of the 3878 adult releases flew straight back to the unconfirmed winter quarters anyway. But it did confirm where juveniles would normally winter, it did confirm that many of the trapped juveniles were still on migration, and it did provide some indication of the distance those birds would travel before their autumnal migratory urges subsided.

Unfortunately, what it did not establish was whether or not the bulk of the transported adults were still on migration as many winter in Holland and adults tend to end their migrations before juveniles. Those that had already completed their migration when caught, were unlikely to re-commence after being dumped in a Swiss city. On the other hand juveniles would be more likely to continue their migrations.

Then comes the puzzle as to why so many juveniles migrated twice as far following displacement compared with the controls, or why they did not maintain the same direction of migration. The organizers played down these two points. Bad enough to try to account for the failure of part one of the hypothesis, without also having to admit to a puzzling part two.

Perdeck's confusing presentation of the results unfortunately disguised the failure of the dual experiment well enough to convince most of the experts that adult starlings could indeed correct for displacement but juveniles could not.

Because of the confusing displays of data, it was virtually impossible to extrapolate the true picture without spending days note taking. In such circumstances, no one could blame students and experts alike for taking the odd short cut, especially when two of Perdeck's diagrams appeared to have simplified the problem.

Perdeck's Figure 8 conveniently represents a combination of Figures 5-7 coincided at one release point (recoveries in March excluded). Certainly much easier than trying to visualize the combined directional results of three sets of data.

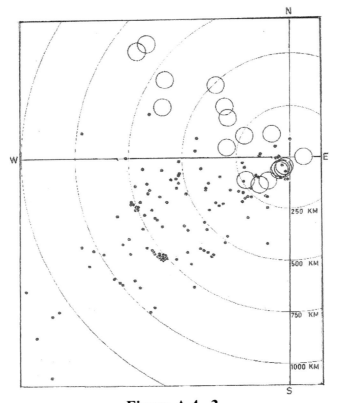

Figure A 4. 3
Perdeck's Figure 8 representing a combination of his
Figures 5-7 coincided at one release point.
O   Adults released without juveniles
•   Juveniles released without adults.
(recoveries in March excluded).
Source: Derived from Perdeck 1958.

One is also presented with a second helpful shortcut. This conveniently
represents Perdeck's Figure 10 coincided at one release point
(recoveries in March excluded).

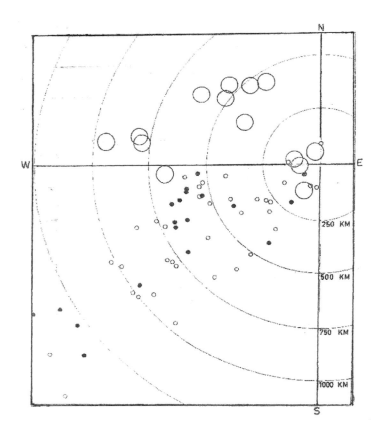

Figure A 4. 4 Mixed transport recoveries.
Perdeck Figure 11 representing his
Figure 10 but coincided at one release point.
O Adults released without juveniles.
• Juveniles with adults.
o Juveniles without adults.
(recoveries in March excluded).
Source: Derived from Perdeck 1958.

Both diagrams display reasonably convincing examples showing most
adults heading northwest towards or actually having arrived in their
assumed winter quarters, and juveniles heading southwest in their innate
southwesterly direction. Both portions of the hypothesis more or less
confirmed.

What is not mentioned is the that 26 circles and 27 dots representing birds caught within 50 kilometers of the three release sites between October and January are not shown either. The absence of geographical outlines is also confusing.

Although data obtained in March when spring migration of adults was already well advanced is sensibly excluded, what is not mentioned in either of these 'coincided to at one release point' diagrams is that data obtained in December to February *is included*; a period long after adult starling autumnal migration has ceased.

If the December to February data is *excluded,* a very different picture emerges as is shown in Figure A 4. 5.

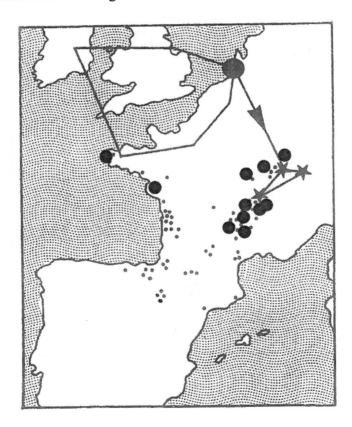

Figure A 4. 5

This diagram represents a combination of Perdeck's Figures 5, 6, 7 *and* 10 (but recoveries in *December to February* as well as March are now excluded).

Note. Figures A 3, 4 and 5 do not include the details of 27 juveniles and 26 adults recaptured within 50 kilometers of the release sites between October and February.

Source: A.C.Perdeck, (1958), redrawn Gerrard (1981a).

Now the true picture emerges. Not one single adult succeeded in reaching its presumed winter quarters before December. By the end of February the following year 10 adults were caught within this wintering resting area. 10 out of the 73 recovered birds, and those 10 taking so long to get there they could have walked most of the way.

But what of the juveniles? The second part of the hypothesis speculated that these inexperienced birds would not be able to compensate for the lateral displacement and would continue on in roughly the same original migratory course (roughly between WSW and SW) for roughly *the same distance* and arrive in the "wrong" wintering area.

Figure A4. 5 reveals that the average distance travelled by many of the recaptured juveniles by the end of November was probably about twice that of the adults and on a different trajectory (as Perdeck admitted).

However significant numbers of juveniles later moved on through the Pyrenees and down into Spain and Portugal. The direction of these longer-distance migrants was different and they had continued for longer, possibly because they had commenced their original migration later. Details are not included in this appendix.

Sadly no one gave much thought to how adult starlings could first determine their displaced geographical position and then work out which way to head in order to rapidly get back on course (or why juveniles could not do so). Nobody seems to have questioned the complex methodology used by Perdeck to illustrate the positions of each recapture either, despite the fact that the above criticisms were first published by the author in book form (1981a) and shortly after in a peer reviewed scientific journal (1981b).

Appendix Five
The original 10 major experimental claims on which our entire understanding of long-distance migratory navigation by birds is based.

The flaws.
In the beginning.
How do migrating birds find their way? This question was first raised more than 2,000 years ago. Could they be using the Sun as a compass?

One. The solar compass.
The details of an experiment that at long last confirmed this belief were first published in German 1949 and in English in 1952 by **G.Kramer**, of the Max Planck Institute, Wilhelmshaven.
1949. '*Über Richtung-stendenzen bei der nächtlichen Zugenruhegekäfigter Vögel*'. Ornithologie als biologische Wissenschaft pp. 269-283.
1952. Publicly announced at the Tenth International Ornithological Congress. '*Eine neue Methode zur Erforschung der Zugorientierung und die bisher damit erzielten Ergebnisse*'.
1952. '*Experiments on Bird Orientation*'. Ibis 94; pp. 265-285.

But massive radar studies in the 1950's showed millions of birds migrated at night even in the absence of the Moon. Could they be using the stars as a compass also?

Two. The stellar compass.
Details of first experiment confirming this discovery were first published in German in 1956 and in English in 1958 by **E.G.Franz (von) Sauer** of Freiburg University.
1956. '*Zugorientierung einer Mönchsgrasmücke(Sylvia a. atricapilla) unter künstlichem Sternhimmel*'. Naturwissenschaften 43: pp. 231-2.
1958. '*Celestial Navigation by Birds*'. Sci. Amer. 199; pp. 42-47.
1960. Publicly announced at the 1960 Cold Spring Harbor Symposium by **E.G.F. & E.Sauer**. '*Star Navigation of Nocturnal Migrating Birds - The 1958 Planetarium Experiments*'.

However, now many migrants were being recorded flying over land beneath solid overcast at night.

Three. Orientation in the absence of Sun or stars.

Details of the first experiment claiming that robins could detect and respond directionally to magnetism inside a building and in the absence of outside cues were first published by **F.W.Merkel** and his assistant **H.G.Fromme** of the University of Frankfurt am Main.

1958. *'Untersuchungen über das Orientierungsvermögen nächtlich ziehender Rotkehlchen (Erithacus rubecula)'*. Naturwissenschaften 45: pp. 499-500.

At this point using Earth's magnetic field to actually navigate from A to B was merely implied, and the question raised by this experimental claim was "Can birds actually use Earth's magnetic field as a directional compass?"

Four. The magnetic compass.

Details of first experiment claiming they might be able to achieve this complex feat was published by **F.W.Merkel** and another of his assistants, **W.Wiltschko**.

1965. *'Magnetismus und Richtungsfinden zugunruhiger Rotkehlchen (Erithacus rubecula')*. Vogelwarte 23: pp. 71-77.

But how can the possession of a magnetic compass tell a migrant bird when to stop? Answer. By detecting the variations in the Earth's magnetic field.

Five. The magnetic intensity and angle of dip compass.

Details of this claim, using the same type of test apparatus used previously by Merkel, was first publicly announced in English by **W.Wiltschko**.

1970. *'The influence of magnetic total intensity and inclination of directions preferred by migrating European Robins'*. NASA.SP 262: pp. 569-578.

The question raised by this claim was how do migrant birds acquire the information required to use a magnetic compass? Could they inherit the skills?

Six. Inherited directional information.

Details of the experiment confirming this suggestion (still using the original type of Merkel test apparatus used in the previous three claims) were published in English by **W.Wiltschko** and **E.Gwinner** (Max Planck Institute and Stanford University).

1974. *'Evidence of an innate magnetic compass in garden warblers'.* Naturwissenshaften 61: pp. 406

The inherited magnetic compass discovery, in turn raised the next question. Different populations of the same species are known to migrate in different directions. The inherited information must therefore vary between groups. Does it?

Seven. Genetic links to migratory direction differences within same species.

Yes, but the details were a long time in coming this time and were not published until 1989 by **A.Helbig, P.Berthold** (both of the Max Planck Institute) and **W.Wiltschko.** But by now a more convenient test rig was being used; the Emlen cage.

1989. *'Migratory orientation of Blackcaps (Sylvia atricapilla): Population-specific shifts of direction during autumn'.* Ethology 82: pp. 307-315.

But if different groups inherit different directional information, crossbreeding between these groups must surely provide the offspring with fresh, different navigational information.

Eight. Genetic basis for migratory directions can be changed by cross-breeding.

This remarkable discovery was made by **P.Berthold, A.Helbig, G.Mohr and U.Querner.**

1992. *'The genetics of bird migration: timing and direction. Genetic basis for migratory directions can be changed by cross-breeding'.* Ibis 134 suppl.1: pp. 35-40. (**P.Berthold** and **A.Helbig**).

1992. '*Rapid microevolution of migratory behaviour in a wild bird species*'. Nature 360: pp. 668-670. (**P.Berthold, A.Helbig, G.Mohr and U.Querner**).

Having solved the entire avian navigation puzzle, the question of what operates the magnetic compass was raised......

Nine. Melatonin essential for working of magnetic compass.
.....and answered by **W.Wiltschko, T. Schneider, H-P.Thalau** and **P.Semm**.
1994. '*Melatonin is crucial for the migratory orientation of Pied Flycatchers (Ficedula hypoleuca)*'. J. E. B. 194: pp. 255-262.

The crucial melatonin discovery was further refined by....

Ten. Magnetic orientation dependant on right eye.
W. and R.Wiltschko and J.Traudt by now all at J W Goethe University, Frankfurt am Main, plus **O.Gunturkin** and **H.Prior** of Ruhr University, Bochum.
2002. '*Lateralization of magnetic compass in orientation in a migratory bird*'. Nature 419: pp. 467-470.

So 10 questions raised one by one in a logical sequence, somehow managed to produce explanations backed by experimental proof, one by one in the same logical sequence. Every one of these claims stems from the same scientific family of German researchers and one man, **W.Wiltschko,** authored/co-authored 6 and worked with authors of 2 others.

This covers all the original basic experimental claims conducted in various test rigs with the exception of experiments by **S.T.Emlen** of Cornell University, which revealed that nocturnal migrants when tested under planetarium skies could not take up the correct orientation - thus confounding the **F.Sauer** claim of a "genetic star map". Instead **Emlen** suggested "a maturation process in which the stellar cues come to be associated with a directional reference system provided by the axis of celestial rotation."
1970. '*Celestial rotation; Its importance in the development of migratory orientation*'. Science 170: pp. 1198-1201.

1975. '*The stellar-orientation system of a migratory bird*'. Scientific American 233:2, pp. 102-111.

A brief outline of the most obvious flaw in each of the 10 papers.

Claim 1. Kramer's Sun compass claim.
Kramer's test cage contained 6 equally spaced windows round the perimeter. He stated that his test subjects would only head towards a window, never a wall. Yet the directional results were computed to the nearest eighth compass point. Phototactic escape responses were not considered.

Claim 2. Sauers' stellar compass claim.
E.G.F.Sauer's planetarium projected only the most prominent of stars (no Moon or planets). The "times" at which he conducted each experiment conveniently coincided with the brightest stars being projected in the direction in which he "expected" the birds to head. Phototactic escape response, (a phenomenon Sauer was well aware of) was responsible for the results that, unsurprisingly have never been replicated.

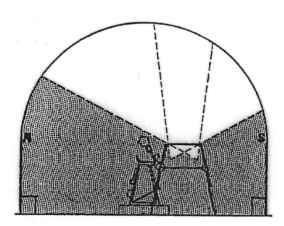

A. 5 Figure 1 a The Sauer planetarium complex.
Indicating the minimum/maximum range of the bird's vision depending on which side of the ring perch was selected by the bird.
Note the position of the projector and the tiny size of the planetarium.
Source: Derived from Sauer (1960).

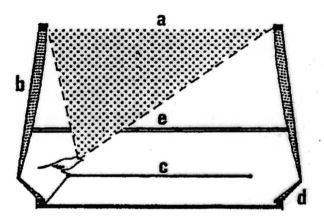

A. 5 Figure 1 b The Sauer planetarium cage.
a. Range of bird's vision.   **b.** Side netting.   **c.** Ring perch.
d. Observer's view hole.   **e.** Plexiglass cover.
Source: Derived from Sauer (1960).

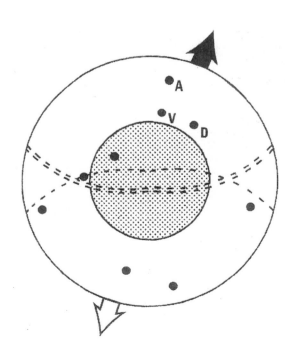

A. 5 Figure 2 The Sauer Figure 4.

The double dotted line represents the visual upper limit and the solid arrow indicates the average directional choice – straight towards the three bright stars.
The single dotted line represents the visual limit when the sky was shifted 180 degrees –thus obscuring Deneb, Vega and Altair –and the hollow arrow, the average directional choice.
Only the 9 first magnitude stars are shown although 2nd, 3rd and 4th magnitude stars were also projected.
The hatched area represents the overhead area more than 50 degrees above the horizon and the outer circle, the visual horizon.
Source: Sauer (1960) adapted Gerrard (1981a).

Claims 3-6. The 4 magnetic compass claims.

These experiments were all conduced using a Merkel cage (A.5 Figure 3). This consisted of an eight-sided circular cage, with 8 sets of radial perches. Whenever a bird alighted on one of the perches, a directional "hit" was registered automatically. These hits were bulked statistically and used to produce the directional claim. However the registration equipment did not indicate which way the bird was facing when it alighted on the perch.

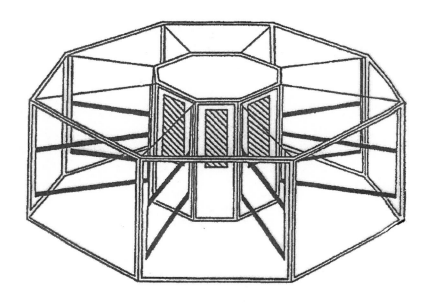

A. 5 Figure 3   One image of a Merkel cage.

Claims 7 -10. The 2 genetic link and 2 magnetic trigger claims.
These used the Emlen test cage, (A.5 Figure 4) which consists of a
funnel with sloping sides and a mesh top to stop the bird escaping. Each
time the bird flutters up (or down) the sides of the funnel, its scratch
marks are recorded. These are bulked statistically. There are many
problems associated with this type of test rig used in over 100 published
claims – and still counting.

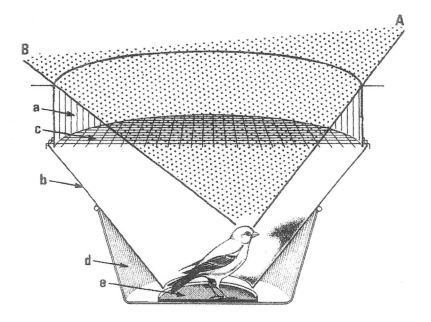

A. 5 Figure 4   The Emlen test cage.
a. opaque circular screen. **b.** blotting paper funnel (later Tipp-Ex).
c. wire screen top.   **d.** two quart pan.   **c.** inky pad (later removed).
The projected lines of vision (A & B) have been added.
Source: S.T.Emlen & J.T.Emlen (1966) Gerrard (1981a).

The most obvious are that the bird cannot see much (of the stars or
whatever) from the bottom of the funnel, so there is no way of checking
which marks represent the "directional" attempt and which represent a
slide back down, and inmates are known to be attracted to imperfections
(such as scratch marks) inside the apparatus.

All the following points have also been noted in publications by researchers using Emlen cages:-
The inmate refuses to become sufficiently agitated to hop about in any direction in total darkness, so light, however diffuse, has to be present. Slightly too much light or unbalanced illumination induces recognisably direct or menotactic phototactic escape responses. Some birds go to sleep standing up and others hang from the wire mesh ceiling if provided. Noise, however minute plays havoc with directional choices. So any test done outdoors is fraught with problems – or open to abuse.

Then comes the problem of registering the directional choices. The original ink pad in the base tray was substituted by Tipp-Ex paper funnels so that the scratch marks could be counted more precisely. But are scratch marks representative of the intended migratory direction, because if so those marks are *never* all in that (assumed) direction (example Tailpiece Figure T1).

The fact that the inmate might be just attempting to escape or is simply scrabbling up the funnel slope to get a better view of the outside world is ignored. This is because many of the marks, when lumped together do indeed *on occasion* seem to point in one (meaningful) direction. And the longer the inmate is left scrabbling about, the more convincing might the data become.

This has been taken to ridiculous extremes. Each directional scratch by a particular inmate is sometimes recorded for 1½ or even 2 hours. The direction of all the scratches is then converted to a mean or average overall direction. The bird is then subjected to up to several further nightly tests, and then the mean of all the means is calculated and set out as one directional dot or triangle on a diagram. The point about attraction to the bird's own scratch marks is never mentioned or joins in the paper funnel rarely considered.

Because the first few hops are very very rarely convincing and because neither are the responses of one bird, long periods of active testing of birds in batches of cages became the norm. This in turn required special methods of statistical analysis. These methods also became more and more bewildering to almost all but the most dedicated. If one selects any current paper linked with an Emlen type cage orientation claim, it is no longer possible to fathom exactly which statistical method was used, as discussed in **Tailpiece**.

Papers always refer readers to previous papers for certain details (the authors would have to write a book otherwise) but all too often that paper, even if accessible to critics, does not provide the answer either, and so on, back eventually to a paper that should have never been published in the first place.

As mentioned in Chapter 1, not one of these 10 claims appears to have been confirmed by independent researchers in a scientific publication and neither have any of the hundreds of subsidiary experimental claims that these 10 stimulated.

Note. Appendix 5 references are excluded from the Bibliography.

Appendix Six

Coded messages and deciphering thereof

As was mentioned in Chapters 3 and 4, scientists have, for whatever reasons, often made use of coded messages. Some deliberately impossible to decipher; a good example being the one Christiaan Huygens sent to the Royal Society in 1675. Some were very risky; as in the case of Isaac Newton's pseudonym. **Jsaacvs Neuutonus**– which was an anagram of **Jeova Sanctus Unus** (One Holy God and thus denying the Holy Trinity) was dangerously close to the Latin version of Newton's name –**Isaacus Neuutonus**.

Some contained brilliantly constructed double meanings (Galileo to Kepler Chapter 3), and very rarely one that contained a second hidden decrypt that only the recipient would be aware of (Wren to Newton). It took the author several years to realise that this Wren cipher held a second meaning and deciphering it took even longer *(Gerrard, Astronomical Minds, Chapter 24, 2007)*.

The Wren basic cipher and the "simple" solution explained in Chapter 4.

OZVCVAYINIXDNCVOCWEDCNMALNABECIRTEWNGRAMHHCCAW

ZEIYEINOIEBIVTXESCIOCPSDEDMNANHSEEPRPIWHDRAEHHXCIF

EZKAVEBIMOXRFCSLCEEDHWMGNNIVEOMREWWERRCSHEPCIP

Read from the right in each line transferring every third letter to a second list. The decrypt relates to 2 (or 3) inventions of Sir Christopher Wren and Robert Hooke.

The following four 46 letter ciphers are all based on the Wren method, but none have second level hidden meanings.

The first is easy to decipher, the second a little more difficult but once mastered, the third, although tricky can be solved. But the fourth is virtually impossible to solve.

No prizes, but if anyone does manage to decipher the fourth, they will deserve one.

First.
SZKNVILIWOXDAMHSMERDUTRCUARTRSNRETECEGPNCAIEVA

Second.
TZORVSAIESXNYMANMPHDUERRWAAARTRRIUEEMGFECYTEFB

Third.
TKANHISLSCROGWSPWETPRSBENMANMMSXEAISSICMIUOINN

Fourth.
EZFMIQTOOZLZMRYTEAXBEALWTLOEAFXTUIUPXBATNFDUQN

BIBLIOGRAPHY

Bairlein, F., et al. 2012, 'Cross-hemisphere migration of a 25g songbird'. *R.S. bio. lett.,* published on line.

Barlow, H.B. and Ostwald, T.J. 1972, 'Pecten of the Pigeon's Eye as an Intra-ocular Eye Shade'. *Nature New Biology,* vol. 236, pp. 88-90.

Berthold, P. (Editor) 1991. *'Orientation in Birds'.* Birkhauser.

Dement'ev, G.P. et al.1954. *'Birds of the Soviet Union',* vol. 6.

Eastwood, E.1967. 'Radar Ornithology'. Methuen.

Emlen, S.T. & Emlen, J. T. 1966, 'A technique for recording migratory orientation of captive birds'. *AUK* vol. 83, pp. 361-367.

Gerrard, E.C. 1981a. Instinctive Navigation of Birds, *Scottish Research Group.*

Gerrard, E.C. 1981b. 'The Perdeck Saga'. *Speculations in Science and Technology* vol. 5, pp. 159-166.

Gerrard, E.C. 1981c. 'Avian migrolution –An introductory attempt at synthesis'. *Scottish Research Group* TXU 73-548.

Holland, R.A. 2014. 'Review. True navigation in birds: from quantum physics to global migration'. *Journal of Zoology,* vol. 293, no.1.

Horridge, A. 2009. 'What does the Honeybee see?' *ANU E Press.*

Kramer, G. 1959.'Recent experiments on bird orientation'. *Ibis,* vol.101, pp, 399-416.

Krebs, J.R. & Partridge, B. 1973. 'Significance of head tilting in the great blue heron'. *Nature New Biology,* vol. 242, pp. 533-535.

Landers, T.J. 2011. 'Dynamics of seasonal movements by a trans-Pacific migrant, the westland petrel'. *The Condor* vol.113, pp. 71-79.

Matthews, G.V.T. 1968. 'Bird Navigation'. *Camb. Uni. Press.*

Myers, M.T. 1964. 'Dawn ascent and re-orientation of Scandinavian thrushes *(Turdus spp.)* migrating at night over the north-eastern Atlantic Ocean'. *Ibis* vol. 106, pp.7-51.

Osman,W.H.1950. 'Pigeons in World War 11'. *Racing* Pigeon *Pub.Co.*

Ozarowska, A et al. 2013. 'A new approach to evaluate multimodal orientation behavior of migratory passerine birds recorded in circular orientation cages'. *J.E.B.* vol. *216.* pp. 4038-4046.

Perdeck, A.C.1958. 'Two types of orientation in migrating starlings *Sturnus vulgaris* and chaffinches *Fringilla coelebs* as revealed by displacement experiments'. *Ardea* vol. 46, pp. 1-37.

Santschi, F. 1911. 'Le mecanisme d'orientation chez les foumis'. *Rev.Suisse Zool.* vol. 19, pp. 117-134.

Schmaljohann, H., Fox, J.W. & Bairlein, F. 2012. 'Phenotypic response to environmental cues, orientation and migration costs in songbirds flying halfway around the world'. *Animal Behaviour,* vol. 84, pp. 623-640.

Shaffer, S.A. et al. 2006. 'Migratory shearwaters integrate oceanic resources across the Pacific Ocean in an endless summer'. *PNAS,* vol. 103, pp, 12799-12802.

Svensson, L. 1970. 'Identification Guide to European Passerines'. *Swedish Mus. Nat. Hist.,* pp. 83-84.

Thorup, K. et al. 2007. 'Evidence for a navigational map stretching across the continental U.S. in a migratory songbird'. *PNAS,* vol. 104, pp.18115-18119.

Thorup, K. & Holland, R.A. 2009. 'The bird GPS – long-range navigation of migrants'. *Journal of Experimental Biology,* vol. 212, pp. 3597-3604.

Williamson, K. 1953. 'Migration into Britain from the north-west, autumn 1952'. *Scot. Nat.,* vol. 65, pp. 69-94.

Lightning Source UK Ltd.
Milton Keynes UK
UKOW04f0512120815

256770UK00002B/32/P